# WORKBOOK to Accompany

# Residential Construction Academy

## Brick and Block Construction

## Robert B. Ham

**THOMSON**
**DELMAR LEARNING**

Australia • Brazil • Canada • Mexico • Singapore • Spain • United Kingdom • United States

Workbook to Accompany
Residential Construction Academy:
Masonry: Brick and Block Construction
Robert B. Ham

**Vice President, Technology and Trades ABU:**
David Garza

**Director of Learning Solutions:**
Sandy Clark

**Managing Editor:**
Larry Main

**Senior Acquisitions Editor:**
James DeVoe

**Development:**
Ohlinger Publishing Services

**Marketing Director:**
Deborah S. Yarnell

**Marketing Manager:**
Kevin Rivenburg

**Marketing Specialist:**
Mark Pierro

**Director of Production:**
Patty Stephan

**Production Manager:**
Stacy Masucci

**Content Project Manager:**
Andrea Majot

**Art Director:**
Bethany Casey

**Editorial Assistant:**
Tom Best

COPYRIGHT © 2008 by Delmar Learning.

Printed in the United States of America
1 2 3 4 5 XX 09 08 07

For more information contact
Thomson Delmar Learning
Executive Woods
5 Maxwell Drive, PO Box 8007,
Clifton Park, NY 12065

Or find us on the World Wide Web at
www.delmarlearning.com

ALL RIGHTS RESERVED. No part of this work covered by the copyright hereon may be reproduced in any form or by any means—graphic, electronic, or mechanical, including photocopying, recording, taping, Web distribution, or information storage and retrieval systems—without the written permission of the publisher.

For permission to use material from the text or product, contact us by
Tel.  (800) 730-2214
Fax  (800) 730-2215
www.thomsonrights.com

**Library of Congress Card Catalog Number:**
2007025085

ISBN 10: 1-4283-2366-X
ISBN 13: 978-1-4283-2366-7

**NOTICE TO THE READER**

Publisher does not warrant or guarantee any of the products described herein or perform any independent analysis in connection with any of the product information contained herein. Publisher does not assume, and expressly disclaims, any obligation to obtain and include information other than that provided to it by the manufacturer.

The reader is expressly warned to consider and adopt all safety precautions that might be indicated by the activities herein and to avoid all potential hazards. By following the instructions contained herein, the reader willingly assumes all risks in connection with such instructions.

The publisher makes no representation or warranties of any kind, including but not limited to, the warranties of fitness for particular purpose or merchantability, nor are any such representations implied with respect to the material set forth herein, and the publisher takes no responsibility with respect to such material. The publisher shall not be liable for any special, consequential, or exemplary damages resulting, in whole or part, from the readers' use of, or reliance upon, this material.

# Table of Contents

Preface .................................................................................................... ix

## CHAPTER 1  Basic Brick Positions and Brick Sizes ........................................ 1
Objectives ................................................................................................ 1
Keywords ................................................................................................. 1
Chapter Review Questions and Exercises ............................................... 1

## CHAPTER 2  Brick Pattern Bonds ................................................................. 5
Objectives ................................................................................................ 5
Keywords ................................................................................................. 5
Chapter Review Questions and Exercises ............................................... 5

## CHAPTER 3  Masonry Hand Tools ................................................................ 9
Objectives ................................................................................................ 9
Keywords ................................................................................................. 9
Chapter Review Questions and Exercises ............................................... 9

## CHAPTER 4  Masonry Construction Equipment ........................................ 13
Objectives .............................................................................................. 13
Keywords ............................................................................................... 13
Chapter Review Questions and Exercises ............................................. 13

## CHAPTER 5  Laying Brick to the Line .................................... 17

Objectives ............................................................................................. 17
Keywords ............................................................................................. 17
Chapter Review Questions and Exercises ........................................... 17

## CHAPTER 6  Constructing Brick Leads .................................... 21

Objectives ............................................................................................. 21
Keywords ............................................................................................. 21
Chapter Review Questions and Exercises ........................................... 21

## CHAPTER 7  Masonry Spacing Scales .................................... 25

Objectives ............................................................................................. 25
Keywords ............................................................................................. 25
Chapter Review Questions and Exercises ........................................... 25

## CHAPTER 8  Masonry Mortars .................................... 29

Objectives ............................................................................................. 29
Keywords ............................................................................................. 29
Chapter Review Questions and Exercises ........................................... 29

## CHAPTER 9  Concrete Masonry Units .................................... 35

Objectives ............................................................................................. 35
Keywords ............................................................................................. 35
Chapter Review Questions and Exercises ........................................... 35

## CHAPTER 10  Laying Block to the Line .................................... 39

Objectives ............................................................................................. 39
Keywords ............................................................................................. 39
Chapter Review Questions and Answers ........................................... 39

## CHAPTER 11 Constructing Block Leads ... 43
Objectives ... 43
Keywords ... 43
Chapter Review Questions and Exercises ... 43

## CHAPTER 12 Estimating Masonry Costs ... 47
Objectives ... 47
Keywords ... 47
Chapter Review Questions and Exercises ... 47

## CHAPTER 13 Residential Foundations ... 51
Objectives ... 51
Keywords ... 51
Chapter Review Questions and Exercises ... 51

## CHAPTER 14 Constructing Water-Resistant Brick Veneer Walls ... 59
Objectives ... 59
Keywords ... 59
Chapter Review Questions and Exercises ... 59

## CHAPTER 15 Anchored Brick Veneer Walls ... 65
Objectives ... 65
Keywords ... 65
Chapter Review Questions and Exercises ... 65

## CHAPTER 16 Composite and Cavity Walls ......... 73

Objectives ..... 73
Keywords ..... 73
Chapter Review Questions and Exercises ..... 73

## CHAPTER 17 Brick Paving ..... 79

Objectives ..... 79
Keywords ..... 79
Chapter Review Questions and Exercises ..... 79

## CHAPTER 18 Steps, Stoops, and Porches ..... 83

Objectives ..... 83
Keywords ..... 83
Chapter Review Questions and Exercises ..... 83

## CHAPTER 19 Piers, Columns, Pilasters, and Chases ..... 87

Objectives ..... 87
Keywords ..... 87
Chapter Review Questions and Exercises ..... 87

## CHAPTER 20 Appliance Chimneys ..... 91

Objectives ..... 91
Keywords ..... 91
Chapter Review Questions and Exercises ..... 91

## CHAPTER 21 Masonry Fireplaces .......... 97
- Objectives .......... 97
- Keywords .......... 97
- Chapter Review Questions and Exercises .......... 97

## CHAPTER 22 Brick Masonry Arches .......... 105
- Objectives .......... 105
- Keywords .......... 105
- Chapter Review Questions and Exercises .......... 105

## CHAPTER 23 Cleaning Brick and Concrete Masonry .......... 109
- Objectives .......... 109
- Keywords .......... 109
- Chapter Review Questions and Exercises .......... 109

## CHAPTER 24 Masonry as a Career .......... 115
- Objectives .......... 115
- Keywords .......... 115
- Chapter Review Questions and Exercises .......... 115

## CHAPTER 25 Safety for Masons .......... 119
- Objectives .......... 119
- Keywords .......... 119
- Chapter Review Questions and Exercises .......... 119

## CHAPTER 26 Working Drawings and Specifications ... 125

**Objectives** ... 125
**Keywords** ... 125
**Chapter Review Questions and Exercises** ... 125

# Preface

## Introduction

Designed to accompany *Residential Construction Academy: Masonry: Brick and Block Construction*, first edition, this workbook extends the core text by providing additional review questions and exercises that will challenge and reinforce the student's comprehension of the concepts presented in *Masonry: Brick and Block Construction*.

## About the Text

The workbook is divided into chapters, with each chapter directly corresponding to a chapter in *Residential Construction Academy: Masonry: Brick and Block Construction*. The chapters contain objectives, quizzes, troubleshooting/quality exercises, and skill-building activities. The quizzes include sections on safety, tools, code regulations and standards, and general knowledge gained from studying the core text and workbook.

## Features of This Workbook

- Outline of major chapter concepts with key terms and definitions
- Additional quizzes and exercises for *Residential Construction Academy: Masonry: Brick and Block Construction*
- Skill-building activities
- Emphasis on troubleshooting skills and quality assurance

## Job Description: Brick Mason

### Skills and Training

Brick masons primarily build walls of clay or shale brick and concrete masonry units, called block, bedding the units individually in mortar to create masonry walls. Brick masons also lay architectural concrete masonry units, glass block, structural tile, and install wall panels made of marble or granite. They sometimes install pre-cast concrete sills, band courses, copings, and caps. A brick mason should be capable of performing basic concrete flatwork such as footings, walkways, patios, porches, and steps for supporting masonry construction.

Brick masonry and cement masonry are separate trades. The latter requires training in placing and finishing architectural concrete—concrete that is permanently exposed to view and that requires special care in forming, placing, and finishing. Neither is a brick mason the same as a stone mason, one trained to cut natural stone and build natural stone walls. To become a brick mason, one trains as an apprentice mason, expecting to serve an apprenticeship period working 40 hours weekly for 3 or 4 years before considered a journey-level worker. A journey-level brick mason should possess the following knowledge and skills.

- Understand working drawings.
- Safely use a variety of hand tools and power equipment.
- Install a variety of masonry materials associated with the trade, including brick, concrete masonry units, architectural concrete masonry units, pre-cast concrete copings and sills, structural tile, glass block, and natural stone wall panels.
- Complete concrete flatwork such as footings upon which masonry walls are built and slab work upon which brick are to be bedded in mortar.
- Erect and use scaffolds in accordance with OSHA standards.
- Know all federal, state, and local regulations governing masonry construction, job-site safety, and environmental issues.
- Although it is not a requirement for a brick mason to possess licensing, formal training and recognition as a journey-level brick mason are available through the department of labor in each state and also through industry training programs such as those of brick masons' labor unions.

## Employment

There is a nationwide shortage of brick masons. The construction industry is one of the United States' fastest growing sectors for employment. As the population increases, so does the demand for homes, schools, churches, retail stores, medical service facilities, product manufacturing facilities, and product distribution warehouses.

- Brick masons work while standing, stooping, or kneeling. As walls are started, masons bend their backs to reach down and install units. As walls are raised, a more erect body position is used. And eventually, they are likely to extend their arms above the shoulders to complete a wall. Such movements require one to have good use of his or her body joints.
- Repeatedly lifting heavier masonry units such as 30-pound blocks for periods of 8 hours and good muscle development are required. Lifting units sometimes with only one hand and manipulating the trowel in the other hand necessitates strong finger gripping to secure the unit and flexible wrist motions for handling the trowel.
- Good eyesight and a sense of proportion are needed for observing the mason's line and aligning pattern bonds uniformly.
- Good physical stamina enables the mason to endure 8-hour days of summer's heat and winter's cold.
- For climbing ladders and working from scaffolds, masons need to overcome any fears of heights.

## Job Outlook

According to the U.S. Department of Labor, Bureau of Labor Statistics, "Job opportunities for brick masons, block masons, and stone masons are expected to be very good through 2014. A large number of masons are expected to retire over the next decade and in some areas there are not enough applicants for the skilled masonry jobs to replace those that are leaving.

Jobs for brick masons, block masons, and stone masons are also expected to increase about as fast as average for all occupations over the 2004–14 period, as population and business growth create a need for new houses, industrial facilities, schools, hospitals, offices, and other structures. Also stimulating demand will be the need to restore a growing stock of old masonry buildings, as well as the increasing use of brick and stone for decorative work on building fronts and in lobbies and foyers. Brick exteriors should remain very popular, reflecting a growing preference for durable exterior materials requiring little maintenance.

Employment of brick masons, block masons, and stone masons, like that of many other construction workers, is sensitive to changes in the economy. When the level of construction activity falls, workers in these trades can experience periods of unemployment."

## *Earnings*

According to the U.S. Department of Labor, Bureau of Labor Statistics, "Median hourly earnings of brick masons and block masons in May 2004 were $20.07. The middle 50 percent earned between $15.34 and $25.20. The lowest 10 percent earned less than $11.68, and the highest 10 percent earned more than $30.43. Median hourly earnings in the two industries employing the largest number of brick masons in 2004 were $22.98 in the nonresidential building construction industry and $19.95 in the foundation, structure, and building exterior contractors industry.

Median hourly earnings of stone masons in 2004 were $16.82. The middle 50 percent earned between $12.74 and $21.45. The lowest 10 percent earned less than $9.97, and the highest 10 percent earned more than $27.23.

Earnings for workers in these trades can be reduced on occasion because poor weather and slowdowns in construction activity limit the time they can work. Apprentices or helpers usually start at about 50 percent of the wage rate paid to experienced workers. Pay increases as apprentices gain experience and learn new skills."

# Chapter 1: Basic Brick Positions and Brick Sizes

## OBJECTIVES

Upon completion of this chapter, you will be able to:
- Identify the six brick positions.
- Explain how extruded brick and wood-mold brick are formed.
- Identify the different sizes of brick available as standard production units.

## Keywords

Cored Brick
Extruded Brick
Frog
Header
Rowlock
Sailor
Shiner
Soldier
Solid Masonry Unit
Stretcher
Wood Mold Brick

## Chapter Review Questions and Exercises

### MULTIPLE CHOICE

1. For a brick to be considered a solid unit rather than a hollow unit, the minimum percentage of its net cross-sectional area considered as material rather than openings must be at least:

    A. 40%.

    B. 50%.

    C. 75%.

    D. 90%.

2. Brick having a length of 7⅝ inches is:

    A. oversize brick.

    B. queen size brick.

    C. standard size brick.

    D. all of the above.

3. Brick having a width of 3½ inches is:

    A. oversize brick.

    B. standard size brick.

    C. utility brick.

    D. all of the above.

1

4. Brick having a face height of 2¾ inches is:

    A. oversize brick.

    B. queen size brick.

    C. utility brick.

    D. both A and B.

5. Brick having a length of 11⅝ inches is:

    A. hollow brick masonry units.

    B. utility brick.

    C. oversize brick.

    D. both A and B.

## SHORT ANSWER

6. What type of brick is processed by forcing wet clay through a die?

   _____
   _____
   _____
   _____

7. What are the holes or openings extending the entire thickness of a brick called?

   _____
   _____
   _____
   _____

8. Relative to the face of the brick, what can be done with a wood-mould brick that cannot usually be done with an extruded brick?

   _____
   _____
   _____
   _____

9. What is a "frog" and how is it typically positioned when bedding brick in mortar?

   _____
   _____
   _____
   _____

## MATCHING

Identify each brick position and write the letter on the line next to each number.

___ 10. header

___ 11. rowlock

___ 12. sailor

___ 13. shiner

___ 14. soldier

___ 15. stretcher

## COMPLETION

Give the dimensions for each of the brick sizes below.

16. Standard size modular brick

   Length (L) = _____ inches
   Width (W) = _____ inches
   Height (H) = _____ inches

17. Engineered or oversize brick

   Length (L) = _____ inches
   Width (W) = _____ inches
   Height (H) = _____ inches

18. Queen size brick

   Length (L) = _____ inches
   Width (W) = _____ inches
   Height (H) = _____ inches

19. Economy, jumbo, or utility brick

   Length (L) = _____ inches
   Width (W) = _____ inches
   Height (H) = _____ inches

20. What would you say to convince those planning new homes to choose a brick façade?

   _____
   _____
   _____
   _____

21. Considering that the initial cost of a brick home is typically higher than that of a home without a brick facade, why is it important for masons to inform the public about the long-term benefits for those choosing to build brick homes?

   _____
   _____
   _____

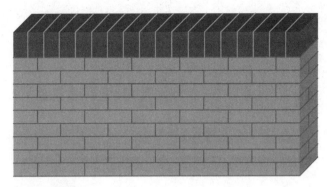

**FIGURE 1-11**

22. Figure 1-11 shows a wall constructed of standard size modular brick. How many bricks laid in the stretcher position are required for each side of each course? _____

23. How many bricks are required to be laid in a rowlock position for the brick capping the top of the wall? _____

24. Using the answers from the two preceding questions, what is the ratio of standard size modular brick laid in the rowlock position to standard size modular brick laid in the stretcher position?

    The ratio is _____ brick in the rowlock position to _____ brick in the stretcher position.

25. Given that the face side of a standard size brick measures 7⅝ × 2¼ inches:

    A. What is the length of a brick including a single ⅜-inch-wide head joint? _____ inches
    B. What is its height including a ⅜-inch-wide bed joint? _____ inches
    C. Referring to the Units of Conversion—Decimal Equivalents of Common Fractions section in the Appendix, what is the answer to question B when the fraction is converted to its decimal equivalent? _____
    D. Given the formula Area(a) or Square Inches = Length (L) × Height (H), what is the area for the face of a brick, including one head joint and one bed joint? _____ square inches
    E. Given that 1 square foot is equal to 144 square inches, how many standard size bricks are laid to be equivalent to 1 square foot? _____

# Chapter 2  Brick Pattern Bonds

**OBJECTIVES**

Upon completion of this chapter, you will be able to:
- Define the term pattern bond.
- Identify the five brick pattern bonds.
- Dry bond each of the five pattern bonds.
- Lay out and build a brick wall for each of the pattern bonds.

## Keywords

American Bond
Bat
Dutch Corner
English Bond
English Corner
Flemish Bond

Flemish Garden Wall Bond
Garden Wall
Pattern Bond
Queen Closure
Running Bond
Screen Wall

Single-Wythe Brick Wall
Snap Header
Stack Bond
Wythe

## Chapter Review Questions and Exercises

### SHORT ANSWER

1. When cutting brick, or at other times when there is a potential for airborne particles causing eye injuries, what personal protection is required?

   _____
   _____
   _____
   _____

2. Is a single-wythe, 4-inch brick wall considered as a structural bearing wall, capable of supporting loads other than its own?

   _____
   _____
   _____
   _____

3. Because there is no overlapping of units between brick courses, how must walls constructed in the stack bond pattern be reinforced?

_____

_____

_____

_____

## COMPLETION

4. The _____ bond is built of alternating courses of stretchers and headers.

5. There is no overlapping of units with the _____ bond.

6. Except for pieces at openings and the ends of walls, brick in the _____ bond are laid in the stretcher position.

7. The _____ bond is built by alternating stretchers and headers in each course.

8. The _____ bond has a course of headers on every fifth, sixth, or seventh course and all other courses are laid in the stretcher position.

9. The bond most often used to build today's brick walls is the _____ bond.

10. Headers used in a single-wythe wall are called _____ or _____ headers.

11. The _____ bond requires horizontal joint reinforcement.

12. The term "common" bond is sometimes used as reference to the _____ bond.

13. A single-wythe brick wall supporting nothing other than its own weight is called _____.

14. Variations of the Flemish bond wall to create decorative patterns are known as _____ _____ bonds.

Identify the brick pattern bonds.

15. _____ or _____

16. _____

17. _____

18. _____

19. _____

20. Assume that you are a masonry contractor and someone intending to use the running bond for the brick façade of a new eighteenth-century American colonial style home asks you to bid on doing the brickwork.

    A. Is the running bond brick pattern a correct representation of nineteenth-century brickwork?

    B. What brick bond pattern(s) could you recommend so that the brick pattern is a correct representation of brickwork typical for the era the house replicates?
    _____
    _____

21. Does it take more time to build a 4-inch single-wythe wall in the Flemish bond or English bond pattern than it does to build the same wall in the running bond pattern? If so, why?
    _____
    _____

22. Assume that you are a masonry contractor and that someone asks how much you will charge them to lay the brick façade of their new home. Why is it important to know the intended brick bond pattern before figuring the total labor cost?
    _____
    _____

23. For someone desiring a stack bond pattern, why would it be a good idea for them to see a sample panel of the selected brick laid in the stack bond pattern?

    Lay out the brick pattern as illustrated in Figure 2-4. Continue the pattern for a third and fourth course.
    _____
    _____

FIGURE 2-4

24. Do the two wythes appear as separate walls or do they appear to be structurally bonded to each other?
_____
_____

25. Would you consider this wall to be two 4-inch brick walls or a single 8-inch brick wall exhibiting greater strength than two separate 4-inch brick walls?
_____
_____

26. Lay out the brick corner as illustrated in Figure 2-5. Have your instructor provide you with needed precut queen closure brick or demonstrate the proper technique for cutting brick by using a mason's hammer and brick set.
_____
_____

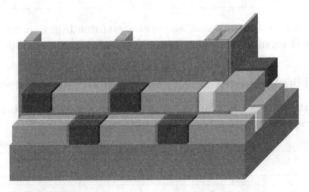

FIGURE 2-5

**Caution:** Wear approved eye protection when cutting brick.

27. Does the façade or face of the 4-inch single-wythe Flemish bond wall with snap headers appear any different from the 8-inch double-wythe Flemish bond wall having full-length brick headers?
_____
_____

28. What are some of the advantages of wall systems having 4-inch single-wythe brick walls anchored to wood or steel frame structural walls?
_____
_____

29. What happens to the bond pattern if the queen closure is omitted and snap headers are separated only by the width of a head joint?
_____
_____

# Chapter 3  Masonry Hand Tools

**OBJECTIVES**

Upon completion of this chapter, you will be able to:

- Identify masonry hand tools.
- Describe available options for specific tools.
- List manufacturers of specific masonry hand tools.
- List safety precautions and care for specific tools.

## Keywords

| | | |
|---|---|---|
| Heel | Plumb | Trowel Shank |
| Level | Toe | Vial |
| Lift | Trowel Blade | |

## Chapter Review Questions and Exercises

### SHORT ANSWER

1. What personal protection equipment is required whenever there is the potential for airborne dust or masonry fragments causing eye injuries?

   _____
   _____
   _____

2. In addition to eye protection, what personal protection equipment should one have when striking chisels?

   _____
   _____
   _____

3. What should be the immediate course of action taken when tools are found to be damaged or otherwise unsafe?

   _____
   _____
   _____

4. Why should one's finger never be put through the hang hole of a mason's level?
_____
_____
_____

5. Because of the risk of dislodging from the wall and becoming a dangerous airborne projectile, what should never be used to secure a tensioned mason's line?
_____
_____
_____

6. What four joint finishes are not recommended for mortar joints of exterior walls that experience severe weather conditions such as fluctuating temperatures resulting in freezing and thawing?
_____
_____
_____
_____

## COMPLETION

7. The _____ _____ _____ is used to apply mortar to brick and block.

8. The three trowel blade patterns are the _____, the _____ _____, and the _____ patterns.

9. Three brands of trowels are _____, _____, and _____.

10. The perimeter of the head of the mason's hammer is _____ shaped, enabling it to make straighter trim cuts on masonry materials.

11. The weight of a mason's hammer typically ranges from _____ ounces to _____ ounces.

12. A _____ jointer forms mortar joints having a concave surface.

13. A _____ jointer creates a recessed bead in the center of the mortar joint.

14. Unlike other mortar joint finishes, the _____ jointer removes mortar from the joint rather than compressing the mortar.

15. The _____ jointer forms a mortar joint having two sloping flat surfaces at right angles to one another.

16. The _____ / _____ jointer duplicates two joint finishes of colonial American brickwork.

17. Used often for stone masonry, the _____ _____ forms a convex joint, one protruding beyond the face of the wall.

18. The _____ _____ is used to align walls perpendicular or parallel to the forces of gravity.

19. The mason's rule or tape measure has markings called _____ for the uniform course spacing of masonry units.

20. A _____ _____ hooks around an outside corner and secures the mason's line under tension.

21. A _____ _____ is a flat, tapered piece of metal that is driven into a head joint to secure a mason's line under tension.

22. A _____ _____ is a flat piece of metal that is clipped onto the mason's line to eliminate the line from moving or sagging.

23. A _____ _____ has a blade that is about 4 inches wide and is struck with a hammer to break brick or block.

24. A _____ chisel is used to remove mortar from mortar joints.

25. A _____ drill or chisel is impacted with a hammer to cut a round hole in masonry.

26. A _____ trowel is a small trowel used to fill mortar joints.

27. A _____ has a narrow blade and is used to pack mortar into mortar joints.

28. A _____ trowel has a rectangular-shaped blade and is used to place or remove mortar in small spaces.

29. A _____ line is made when a chalk-coated, cotton string line is tensioned, lifted along its length, and released so that the chalk is transferred from the string line to the surface below it.

Identify the Parts of a Brick Mason's Trowel

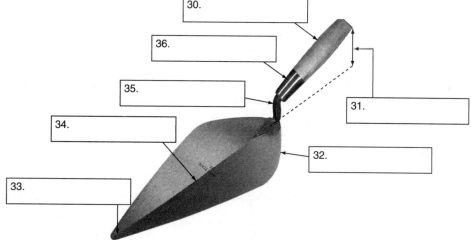

37. Assume that you are at your local masonry tool supplier and intend to buy a new mason's trowel. You know the brand, blade length, blade style, and the type of handle that you want. Why might it be beneficial for you to hold several of these identical trowels, one at a time as if intending to use it, before selecting a single trowel?

_____

_____

_____

38. Because many masons use identical-looking tools, why should one engrave or otherwise identify their tools for easy recognition?

39. Why should a mason consider bringing duplicate or spare tools and equipment to the job site?

40. Assume that you are a masonry contractor and that you have provided your employees with eye protection, yet some of them carelessly expose themselves to flying debris when cutting brick and block using the mason's hammer and chisel. Is there a safety standard requiring workers to wear eye protection where operations present potential eye injury? If so, how do you suggest having your employees comply with such a safety standard?

41. Obtain several spirit mason's levels from the instructor and test each one for accuracy. Referring to the Procedures section in Chapter 3, "Checking the Accuracy of a Mason's Spirit Level," examine all vials of each level. Tag all levels whose level or plumb vials show inconsistent results. Have the instructor verify your findings.

# Chapter 4 — Masonry Construction Equipment

## OBJECTIVES

Upon completion of this chapter, you will be able to:

- Identify manually operated and power equipment used in the masonry construction industry.
- Discuss factors to consider when selecting specific types of equipment.
- List safety precautions and care for specific equipment.

## Keywords

Dense Industrial 65
Point Up

Sawn Planking
Scaffold Buck

Supported Scaffolds
Suspension Scaffolds

## Chapter Review Questions and Exercises

### SHORT ANSWER

1. In addition to eye protection, what personal protection equipment should one have when performing sawing or grinding operations, using noise-generating power tools having the potential to create airborne fragments?

   _____
   _____
   _____
   _____

2. For one operating a mortar mixer, what personal protection equipment should one have to meet OSHA standards?

   _____
   _____
   _____
   _____

3. What job-site protection is required to protect operators of power tools such as electric motor-powered mixers, saws, drills, and other equipment and machinery from potential electrical shocks or electrocution?

4. What training must employees receive before erecting, disassembling, or accessing scaffolds?

5. What precautions should be taken to protect those who are susceptible to looking directly into laser light beams generated by laser levels?

## COMPLETION

6. Mortar boards typically have a _____-inch to _____-inch square shape.
7. Mortar is mixed manually in a mortar _____.
8. The two large holes in the blade of a mortar _____ improve the mixing of mortar ingredients and reduce the physical effort of mixing.
9. Brick and block can be transported on _____ and _____ wheelbarrows.
10. Contractor-grade wheelbarrows typically have _____ cubic-foot tray capacities.
11. Brick _____ permit the single-handed carrying of several brick at once.
12. Mortar is mixed mechanically in gasoline-engine or electric motor powered _____ _____.
13. List two types of gasoline engine or electric motor powered masonry saws.

14. The electric-powered _____ grinder is used for removing mortar joints.
15. A _____ _____ works on the principle of mechanical leverage to break solid brick pavers.
16. A grout _____ is filled with mortar and manually squeezed, permitting mortar to exit its tip for filling joints with mortar.

CHAPTER 4  *Masonry Construction Equipment*  **15**

17. A powered _____ gun dispenses mortar as mortar is fed mechanically through its tip.

18. Temporary work platforms designed to accommodate both workers and materials are called _____.

19. Give two classifications of scaffolds.
    _____
    _____

20. Tubular steel-leg scaffold end frames rest on _____ plates and are leveled with _____ jacks.

21. Wood scaffold planking must meet the industry standard known as _____ _____.

22. Preventing scaffold boards from splitting are either metal plank _____ or drilled tie _____.

23. Why isn't construction-grade lumber used for floor joists suitable for scaffold planking?
    _____
    _____

24. A rope pulley used to hoist materials is known as a _____ wheel or a _____ wheel.

25. The two classifications of builders' levels are glass _____ lens levels and rotary _____ light levels.

26. Assume that you are a masonry contractor and that a new employee informs you that he has been properly trained to use the types of equipment and machinery that you have. What should you do, presume that the new employee is properly trained and permit him to use the equipment and machinery or evaluate his knowledge and understanding before allowing him to use it?
    _____
    _____
    _____
    _____
    _____
    _____

27. How does it benefit employers to require employees to comply with mandatory safety standards?
    _____
    _____
    _____
    _____
    _____
    _____

28. What should a worker do and who should she notify upon realizing that a piece of power equipment or machinery at a job site is being operated without inline ground fault circuit protection or without mandatory personal protection from electrical shock or fatal electrocution for those operating power equipment on construction sites?

29. Assume that you are a masonry contractor working in a neighborhood development with inhabitants in nearby homes. Out of respect for and courtesy to the nearby residents, are there times that you should refrain from using noisy equipment and machinery? What impact can such a decision have on your reputation as a masonry contractor? Are there ever governing regulations restricting the time of day during which noisy equipment and machinery can be used? If so, what are some of the influencing factors restricting their usage?

# Chapter 5 Laying Brick to the Line

### OBJECTIVES

Upon completion of this chapter, you will be able to:
- Lay out a brick wall in the running bond pattern.
- Demonstrate options for placing a cut brick in a wall.
- List the four procedures performed for laying every brick.
- Demonstrate procedures for hanging a line and twigging a line.
- Lay brick to the line in the running bond pattern.

## Keywords

Closure Brick
Crowding the Line
Dry Bonding
Facing the Brick

Hanging the Line
Holding Bond
Layout
Lipping

Racking
Raising the Line
Set-Back
Twigging the Line

## Chapter Review Questions and Exercises

### SHORT ANSWER

1. Why should eye protection be worn when pulling lines to tension, working near or in the presence of tensioned lines, or removing tensioned lines from walls?
   _____
   _____
   _____

2. Why should twigs be removed from lines before repositioning lines to successive courses, tensioning, or removing lines from walls?
   _____
   _____
   _____

3. Why is it not recommended that you use your eyes to sight the length of a tensioned, twigged line from one end to the other to confirm a straight and level line?
   _____
   _____
   _____

4. What potential hazards are you exposed to while laying brick to a line that necessitates wearing eye protection?

5. What hazards do laser levels present and what precautions must you take when in the presence of laser light leveling equipment?

6. What is the recommended minimum width for mortar joints?

7. What is the recommended maximum width for mortar joints?

8. To maintain level and plumb brick coursing, what relationship does the mason observe between the line and each brick that is laid to the line?

9. What choices does a mason have for securing a tensioned mason's line to leads at opposite ends of a wall?

10. What is sometimes "clipped" onto a mason's line to prevent the weight of the line from causing it to sag?

11. What is used to lay out brick courses uniformly at their intended spacing?

# CHAPTER 5  Laying Brick to the Line

## COMPLETION

12. The process of establishing brick arrangement and head joint size for a course of brick is called _____.

13. The process of selecting a bed joint width permitting a wall to be built to a specific height is called _____.

14. Attaching a mason's line to the leads is referred to as _____ the _____.

15. Three purposes for twigging the line are _____, _____, and _____.

16. Aligning the bottom edge of each brick's face with the top edges of the brick faces below it is known as _____ the brick.

17. A condition observed when the brick unintentionally extends beyond the face of the brick below it is known as _____.

18. A condition exhibited when a brick is unintentionally laid in back of the wall line is known as _____.

19. A brick laid to the line is considered to be aligned plumb when a space no less than _____ inch(es) or more than _____ inch(es) is exhibited between the line and the face of the brick.

20. Having brick too close to the mason's line is referred to as _____ the _____.

21. The last brick to be laid in a course is referred to as the _____ brick.

22. The four operations performed when laying a brick to the line in order of the attention given them are _____, _____, _____, and _____.

23. Although a wall may meet all building code requirements yet exhibit undesirable conditions such as a bond alignment that is not plumb, poorly faced brick, mortar joints that are noticeably not uniform in size, and brick courses that are noticeably not aligned level, why is it important for a masonry contractor to not allow the masons work to exhibit such inferior quality? Over a period of time, what reputation does one receive for completing work below the standards of what is considered professional quality?

_____

_____

_____

_____

24. Although it may be considered a nonproductive task, taking time that otherwise would permit laying brick, what benefits are there to dry bonding the first course of brick for a wall?

_____

_____

_____

_____

25. Assume that you are a masonry contractor and the brick that your masonry crew is to lay on a house have highly noticeable irregular face profiles and differ in length by as much as ½ inch. These are intentional brick variations intended by the brick producer to replicate "old-world" or colonial-period, hand-made brick. Before beginning the brick façade for the home, what can you do to avoid a confrontation with the homeowners who may be disappointed with the appearance of the irregular-shaped brick and the joint size variations that are evident because of them?

26. CAUTION! This exercise requires cut brick. Have your instructor provide the necessary pieces or demonstrate to you proper and safe procedures for using the mason's hammer to cut brick. Wear eye protection and hand protection when cutting brick with a hammer.

    Using standard size modular brick and allowing for ⅜-inch-wide head joints between them, lay out a brick wall 46 inches long, similar to the illustration shown in Chapter 5, Procedures Section 2, "Making Allowances for Walls Where Brick Needs Cutting." Have your instructor supply you with the necessary cut brick. Lay out two courses as shown in Procedures 2, Step C and illustrated in Figure C. Lay out another wall two courses as shown in the Alternate Procedures 2, Step C and illustrated in Figure C.

27. Which method do you prefer, that demonstrated in Procedure 2, Step C or the alternate method for Procedure 2, Step C?

28. What is the advantage of laying the combination of the approximately 6-inch-long piece and the half brick at the right end of the wall on the second course rather than the small, approximately 2-inch-long piece at the end of the second course?

# Chapter 6: Constructing Brick Leads

## OBJECTIVES

Upon completion of this chapter, you will be able to:

- Lay out and construct an outside corner.
- Lay out and construct an inside corner.
- Lay out and construct 4-inch, 8-inch, and 12-inch brick jambs.
- List precautions taken when brick toothing.
- Demonstrate procedures for setting a corner pole.

## Keywords

Brick Jamb
Checking the Range
Corner of the Lead
Double-Wythe Wall
Mortar Bridgings
Mortar Protrusions
Quoined Corner
Rack of the Lead
Tail of the Lead
Toothing

## Chapter Review Questions and Exercises

### SHORT ANSWER

1. What potential hazards should one be protected from while working on scaffolds to build outside corners and jambs?

   _____
   _____
   _____
   _____

2. Describe the test for determining when mortar joints should be tooled.

   _____
   _____
   _____
   _____

3. What is recommended as the maximum projection beyond the face of the wall for cored brick creating quoined corners?

4. What is recommended as the maximum projection beyond the face of the wall for solid brick creating quoined corners?

5. What tool is used to align leads level, plumb, and straight?

6. What is held the length of a wall to "check the range" between leads at opposite ends of a wall?

## COMPLETION

7. A _____ is that part of a wall built first to which a mason's line is attached for building the rest of the wall.

8. Brick corners are classified as being either _____ or _____, depending on which side of it is accurately aligned.

9. A brick _____ is a type of lead that may be constructed next to window or door openings.

10. Without toothing the corner, a brick corner having six brick on each side of the first course can be built to a height of _____ courses.

11. Confirming the alignment of a wall to be straight from one end to another is called _____ the _____.

12. The stepped-back end of a lead is called the _____ of the lead.

13. Holding the straight side of a mason's level diagonally against the brick of each course at the tail of the lead is called checking the _____ of the lead.

14. The actual width of an 8-inch brick jamb is equivalent to the _____ of the brick selected to build the wall.

15. Temporarily laying half-brick at the tail of a lead and continuing to build the lead above the height normally restricted by its racked back design is called _____ the corner.

16. Eliminating the construction of leads is sometimes accomplished by temporarily erecting straight, metal, or wood _____ _____ to which a mason's line is attached for laying brick to the line.

17. The face of a wall is considered to be aligned _____ when it is parallel with the force of gravity.

18. The top of a wall is considered to be aligned _____ when it is perpendicular to the force of gravity.

19. What problems can arise from waiting too long to strike or tool the mortar joints of a brick corner lead?

_____
_____
_____
_____

20. What should two masons who are simultaneously building separate leads at opposite ends of a wall decide to do to ensure that the wall to be built between the leads has uniform bed joint widths?

_____
_____
_____
_____

21. Other than wasting mortar, why should a mason apply no more mortar than necessary when building 8-inch brick jambs?

_____
_____
_____
_____

22. Where else can similar problems occur when building corner leads no more than 1 inch from back-up walls such as structural wall house framing?

_____
_____
_____
_____

# Chapter 7  Masonry Spacing Scales

## OBJECTIVES

Upon completion of this chapter, you will be able to:

- Identify and give the application for the three sets of spacing scales.
- Lay out a wall to a specific height by using the spacing scales.
- Lay out a brick rowlock by using masonry scales.
- Discuss factors influencing the selection of scales.

## Keywords

Mason's Scales             Modular Masonry Construction             Modular Masonry Unit

## Chapter Review Questions and Exercises

### SHORT ANSWER

1. What is the recommended minimum width for bed joints?

   _____
   _____
   _____
   _____

2. What is the recommended maximum width for bed joints?

   _____
   _____
   _____
   _____

3. What inch spacing does modular masonry construction permit adjacent wythes of all modular masonry materials to be at the same level?

   _____
   _____
   _____
   _____

4. What are the three masonry scales called?

5. Besides on 6-foot folding rules, on what other measuring device can masonry scales be found?

## COMPLETION

6. By using the mason's scales, you can ensure _____ course spacing.

7. A total of _____ standard-size brick spacing scales are represented on the masons' spacing rule.

8. Two factors considered when selecting a spacing scale are _____ and _____.

9. The corresponding red numbers beside the scales on the standard-size and oversize brick spacing scales represent the number of _____.

10. Letters _____ through _____ represent the scales on the oversize brick spacing rule.

11. Composite and cavity wall constructions rely on the _____ spacing scales.

12. Modular scaling permits the combination of any two different modular masonry materials being at the same level every _____ inches in height.

13. Suppose that while constructing a modular masonry composite wall, having 2 courses of cement block and 6 courses of modular standard brick spaced at 16-inch intervals, you discover that the bed joints between the brick courses are very narrow, less than the recommended ¼ inch minimum width. What course of action should you take: continue building the wall or consult the person in charge of its construction?

14. What effects could there be to a wall having bed joints whose widths are less than recommended?

15. Assume that you are a masonry contractor and that because of the different heights of some windows in a home, brick coursing permitting full face height of brick cannot be attained to accommodate brick sills below all windows or level coursing with the top of all windows. What can you do to permit the desired spacing below such windows, where brick rowlock sills are intended?

_____
_____
_____
_____

16. What options do you have to ensure level brick coursing above the window?

_____
_____
_____
_____

17. What happens if different masons working on the same project are not building leads having identical course spacing?

_____
_____
_____
_____

18. What circumstances can you think of where it may be desirable to build leads at opposite ends of a wall at different course spacing?

_____
_____
_____
_____

19. Give the modular scale number used as a reference and the number of courses for every 16-inch interval of wall height for the following modular masonry units.

   A. Concrete masonry units are laid on modular scale _____, having _____ courses every 16 inches of wall height.

   B. Facing tile are laid on modular scale _____, having _____ courses every 16 inches of wall height.

   C. Economy brick are laid on modular scale _____, having _____ courses every 16 inches of wall height.

   D. Oversize brick are laid on modular scale _____, having _____ courses every 16 inches of wall height.

   E. Standard size brick are laid on modular scale _____, having _____ courses every 16 inches of wall height.

   F. Roman brick are laid on modular scale _____, having _____ courses every 16 inches of wall height.

20. Refer to the following illustration to answer the questions related to interpreting the readings for brick coursing on a brick mason's spacing rule.

   A. Interpret the reading of the top line, "521."
   _____

   B. Interpret the reading of the middle line, "322."
   _____

   C. Interpret the reading of the bottom line, "720."
   _____

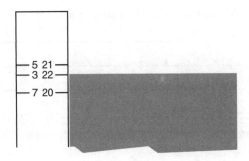

Referring to the illustration, put into your own words three different heights for which the top course could be in relationship to the readings shown on the brick spacing scale.

   D. _____
   E. _____
   F. _____

   G. Other than observing the information provided on the illustrated brick mason's spacing scale and its relationship to the top of the wall beside which it is placed, what additional information is needed to correctly interpret the readings that the ruler shows?
   _____

# Chapter 8  Masonry Mortars

## OBJECTIVES

Upon completion of this chapter, you will be able to:

- List the ingredients of masonry mortars.
- Identify the types of masonry cements and their recommended uses.
- List additives contained in some masonry cements.
- Describe the procedures for mixing mortar manually and with a power mixer.
- List procedures for maximizing the intended performance of mortars.
- Describe the differences between mortars used for new construction and mortars used for tuckpointing the joints of older and historical brick walls.
- Describe potential problems associated with mortars.
- List types of specialty mortars.
- Describe masonry grout operations.

## Keywords

Accelerators
Admixtures
Autogeneous Healing
Bond Strength
Cold Weather Construction
Elasticity
Flexural Strength

Grout
Hot Weather Construction
Masonry Cement
Mortar
Mortar Cement
Pigments
Plasticizers

Proprietary Masonry Cements
Retarders
Retempering
Tensile Strength
Water Retention
Workability

## Chapter Review Questions and Exercises

### SHORT ANSWER

1. What personal protection equipment must one have when exposed to the potential hazards of airborne dry cement dust and dry sand?

   _____
   _____
   _____

2. What personal protection equipment must one have when exposed to the potential hazards of mixed mortar spatters entering the eyes?

3. What are the potential risks for breathing air contaminated with cement dust, sand, or fiberglass, which are ingredients in some types of cements?

4. What are the three categories of cement materials found in masonry mortars?

5. What are three types of mortar cements typically used for new construction?

6. What are three types of masonry cements typically used for new construction?

7. Because no single mortar type is best for all purposes, what advice does the Brick Industry Association give one when choosing the strength of a mortar?

8. What actions must be taken in the presence of cold weather construction to protect construction materials and walls under construction from low temperatures?

9. List four recommendations that the Brick Industry Association gives to prevent compromising the strength of masonry work during hot weather.

10. What is the recommended slump for grout that is to be consolidated within masonry walls for strengthening them?

11. What tool is used to mix mortar ingredients manually?

12. What hand-held tool is typically used to apply surface bonding cements?

13. List three types of power equipment that may be used to consolidate grout that is placed in masonry walls.

## COMPLETION

14. A mixture of masonry cement or mortar cement, sand, and water used to bond masonry units is called _____.

15. The ingredient accounting for the durability and high strength of today's mortars is _____ _____.

16. The ingredient accounting for the workability and water retention of mortar is _____.

17. Small voids and hairline cracks in masonry mortar joints are sealed by a process known as _____ _____, where water and carbon dioxide react with the lime in mortar to form calcium carbonate.

18. A measure of a material's ability to withstand stretching forces is called _____ strength.

19. A measure of a material's ability to withstand bending forces is called _____ strength.

20. Type _____ mortars have maximum compressive strength, but are not very workable.

21. Type _____ mortars exhibit excellent tensile strength and maximum flexural strength and are recommended for masonry walls at or below grade.

22. Type _____ mortars are recommended for most exterior walls above grade including those exposed to severe weather.

23. Types _____ and _____ mortars have limited applications but are desirable for restoration of older brickwork.

24. Higher air content of mortars increases both _____ and _____
_____.

25. Higher air content of mortars decreases both _____ _____
and _____ _____.

26. Ingredients added to mortars for specific purposes are called _____
_____.

27. Ingredients called _____ may be added to masonry cements intended for use when temperatures are expected to drop below freezing for the purpose of reducing the time required for the mortar to set.

28. Ingredients known as _____ may be added to mortar intended for use in hot weather to delay the time for the mortar to harden.

29. Pigments made from synthetic _____ _____ are used in masonry cement because they provide more consistent colors.

30. Ingredients in masonry cement called _____ make it easier to spread and better adhere to the trowel and masonry units.

31. Sand is added to masonry cement to help control _____.

32. White mortar relies on small, sand-size particles resulting from the grinding of _____, _____, or _____.

33. Water containing _____ matter weakens the mortar.

34. For mixing ingredients to make mortar, the Brick Industry Association recommends a ratio of not less than _____ parts sand or more than _____ parts sand to 1 part mortar cement or masonry cement.

35. Adding water to mortar to replace water lost by evaporation is known as _____.

36. Adding water to mortar because of evaporation reduces _____ strength but improves _____ strength.

37. A term that describes mortar intended for repairing mortar joints in which the mortar is allowed to set one or two hours before being used is called _____ mortar.

38. Specialty mortars designed for laying glass block, tile, and marble contain additives such as _____ and _____.

39. Cold temperatures are defined as either air or material temperatures below _____ °F.

40. Assume that you are a masonry contractor and upon arriving at a new job site where you are to build a cement block wall foundation that is below grade level, you discover that the builder is providing type N masonry cement rather than type S masonry cement. Knowing that the building code does not permit using type N masonry cement for foundation walls below grade, what course of action do you recommend taking?

    a. building the walls using type N masonry cement because it is what the builder provided

    b. contacting the builder before beginning the job, although it may mean a loss of time for you and your crew

    _____

    _____

    _____

    _____

41. What measures can a mason take to ensure that each batch of mortar mixed contains equal volumes of sand?

42. Assume that you are a masonry contractor and the one responsible for mixing the mortar in the power-operated mixer has a habit of allowing the mortar to mix much longer than recommended while attending to other tasks. Besides additional costs of operating the mixer, are there other concerns of which that person should be aware?

43. The average compressive strength rating for each of the types of mortars varies. Compressive strength ratings are measured in pounds per square inch (psi). It is a measure of the maximum force or load that an area measuring 1 inch by 1 inch, typically referred to as "one square inch," can support. Material failures are a result if forces or loads heavier than the rating of the materials are imposed. For mortar, it is a rating of the compressive strength of one square inch of mortar into which a brick is bedded.

44. Given three choices, 750 psi, 1800 psi, and 2500 psi, which best describes the compressive strength rating for each of these three types of mortar?

    Type M has a compressive strength rating of _____ psi.

    Type S has a compressive strength rating of _____ psi.

    Type N has a compressive strength rating of _____ psi.

    Assuming the surface area of a standard size or oversize brick measures 3½ × 7½ inches, what is the total surface area of one of these brick sizes?

# Chapter 9  Concrete Masonry Units

**OBJECTIVES**

Upon completion of this chapter, you will be able to:

- Identify types of concrete masonry units.
- Identify the sizes of concrete masonry units.
- List the ingredients of concrete masonry units.

## Keywords

Anchored Veneer
Architectural CMUs
Autoclaved CMUs
Exposed Aggregate CMUs
Fluted CMUs
Glazed CMUs
Ground Face CMUs
Heavyweight CMUs
Hollow Unit
Lightweight CMUs
Solid Unit
Sound-Absorbing CMUs
Split-Face CMUs
Stone-Face CMUs
Structural Load

## Chapter Review Questions and Exercises

### SHORT ANSWER

1. Because CMUs are made of a combination of Portland cement, water, and various fine aggregates including sand, crushed limestone, and lightweight aggregates, before using a masonry saw to cut block, what precautions should one take when exposed to such airborne dusts?

   _____
   _____
   _____

2. How does the industry define a "hollow masonry unit"?

   _____
   _____
   _____

3. List the nominal sizes of the five CMUs typically available from block suppliers.

   _____
   _____
   _____

4. What two sources generally dictate the size of CMUs required for a specific job?
   _____
   _____

## COMPLETION

5. The hollow spaces or openings in concrete masonry units are called _____.

6. A hollow concrete masonry unit is one in which the open cells account for more than _____ percent of the cross-sectional area of the unit.

7. Concrete masonry units containing expanded shale and slate are referred to as _____ _____ units.

8. List three advantages of using block that contain expanded shale or slate.
   _____
   _____
   _____

9. The weight of the building materials, occupants, furnishings, and snow are examples of _____ loads.

10. The abbreviation _____ is used by architects and engineers when referring to concrete masonry units.

11. Structural concrete masonry units having color and/or texture are referred to as _____ concrete masonry units.

12. A _____-_____ block is made by mechanically splitting a larger block to create units having a rough, quarried-stone appearance.

13. Concrete masonry units having polished, smooth faces to expose the natural colors of the aggregates are called _____ _____ units.

14. Concrete masonry units designed to reduce sound transmission are called _____-_____ units.

15. The *size* of a concrete masonry unit refers to its approximate _____.

16. The standard length of a concrete masonry unit is _____ inches.

17. The standard height of a concrete masonry unit is _____ inches.

18. The term _____ size is used to express the approximate size of a concrete masonry unit.

19. For modular masonry construction, the dimensions of concrete masonry units are intended to accommodate _____-inch-wide mortar joints.

20. Assume that you are a masonry contractor. What precautions would you require your employees to comply with to prevent chipping and scarring the faces of ground face, glazed, and sound-absorbing units as they are handled and bedded to build walls?
    _____
    _____
    _____

21. Assume that you are a masonry contractor approached by a homeowner to build CMU walls for a new home's foundation that will permanently be exposed above grade level. They are unaware of the variety of architectural CMUs that are available. What information can you give them to consider as options for plain CMU walls and architectural CMUs that may add value to their home while improving the foundation's quality and appearance?

22. How can taking the time to make them aware of these options improve the reputation of your masonry business?

    The dimensions of an 8-inch CMU are

    - Length (L) = 15⅝ inches
    - Width (W) = 7⅝ inches
    - Height (H) = 7⅝ inches

23. Using the mathematical formula Area (A) = Length (L) × Width (W), calculate the CMU's cross-sectional area. Refer to Appendix A, Decimal and Metric Equivalents of Common Fractions if decimals rather than fractions are used to calculate the answers.

24. Using the mathematical formula Volume (V) = Length (L) × Width (W) × Height (H), calculate the CMU's volume. Refer to Appendix A, Decimal and Metric Equivalents of Common Fractions if decimals rather than fractions are used to calculate the answers.

25. If each of the cells measures 5 inches × 4¾ inches, calculate the cross-sectional area of one cell. Refer to Appendix A, Decimal and Metric Equivalents of Common Fractions if decimals rather than fractions are used to calculate the answers.

26. What is the cross-sectional area of the two cells combined?

27. For a CMU having open cells, net cross-sectional area is equivalent to gross cross-sectional area minus the area of the open cells. Calculate the net cross-sectional area for the 8-inch CMU.

28. Referring to the overall or gross cross-sectional area of the block as calculated in Problem #1, and the net cross-sectional area calculated in Problem #4, what percentage does the net cross-sectional area of the block represent as compared to its gross cross-sectional area?

29. According to the definitions for "hollow" and "solid" masonry units, is this 8-inch CMU considered to be a hollow masonry unit or a solid masonry unit? Why?

# Chapter 10 Laying Block to the Line

**OBJECTIVES**

Upon completion of this chapter, you will be able to:

- Lay out a block wall in the running bond pattern.
- Demonstrate options for placing a cut block in a wall.
- List the four procedures performed for laying every block.
- Demonstrate procedures for hanging a line and twigging a line.
- Lay block to the line in the running bond pattern.

## Keywords

CMU
Face Shell Spreading
Facing the Block
Hanging the Line
Twigging the Line

## Chapter Review Questions and Answers

### SHORT ANSWER

1. Why should eye protection be worn when pulling lines to tension, working near or in the presence of tensioned lines, or removing tensioned lines from walls?
   _____
   _____
   _____

2. Why should twigs be removed from lines before repositioning lines to successive courses, tensioning, or removing lines from walls?
   _____
   _____
   _____

3. Why is it not recommended to use one's eyes to sight the length of a tensioned, twigged line from one end to the other to confirm a straight and level line?
   _____
   _____
   _____

4. What potential hazards is one exposed to while laying block to a line that necessitates wearing eye protection?

5. Because CMUs are made of a combination of Portland cement, water, and various fine aggregates including sand, crushed limestone, and lightweight aggregates, what precautions should one take before using a masonry saw to cut block?

6. For block weighing more than 30 pounds or in any situation where it may be difficult for one mason to bed a block in mortar, what is recommended?

7. What type of gloves may one wish to consider wearing when handling block?

8. What is the recommended minimum width for mortar joints?

9. What is the recommended maximum width for mortar joints?

10. What is the intended width for both head joints and bed joints if modular masonry construction is intended, meaning a combined block and joint spacing 16 inches long and an 8-inch course height?

11. Describe the test for determining when mortar joints should be tooled.

12. To maintain level and plumb block coursing, what does the mason observe at the top of each course when laying block to the line?

13. What choices does a mason have for securing a tensioned mason's line to leads at opposite ends of a wall?

14. What is sometimes "clipped" onto a mason's line to prevent longer length tensioned lines from sagging, lower near the middle rather than the straight line intended?

15. What measuring devices can be used to lay out block spacing along the length of a wall or the vertical course spacing?

## COMPLETION

16. For modular masonry construction, the combined length of a block and a head joint is _____ inches.

17. For modular masonry construction, the combined height of a block and a bed joint is _____ inches.

18. Give four purposes for twigging the line:

19. Applying mortar along each side of the top of block without applying mortar to the cross webs is referred to as _____-_____ _____.

20. A uniform half-lap, running bond can be maintained by _____ a block's center cross web above the adjoining ends of the two block on the course below it.

21. Aligning the bottom edge of each block as it is laid with the top edges of the block faces below it is referred to as _____ the block.

22. Block should be leveled by tapping along the _____-line of its length to prevent unintentionally tilting the face of the block.

23. To maintain a plumb wall, a spacing equivalent to the thickness of a _____ coin should be provided between each block and the mason's line.

24. Mortar joints should be no more than _____ inches and no less than _____ inches to ensure joints having intended strengths.

25. To preserve a block's structural integrity or reliability, use a masonry _____ rather than a hammer and chisel to cut block.

26. Assume that you are a masonry contractor constructing block walls for a homeowner's new garage. You overhear the homeowner having a conversation about how you are applying mortar bed joints on top of the block courses. He is concerned that you are not applying mortar to the cross webs of the block, all of which have a single center web and noticeable "ears" at both ends rather than square end block. What should you do?

_____
_____
_____
_____

27. Assuming the overall length of the wall to be 82 inches as shown in Figure 10-1, what should be the length of the cut block indicated by the arrow?

_____
_____

**FIGURE 10-1**

28. Assuming mortar joints to be ⅜ inch wide, give the correct measurements as indicated for the CMU wall in the following illustration.

_____
_____

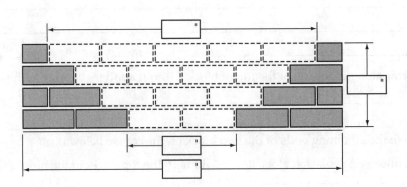

# Chapter 11 Constructing Block Leads

## OBJECTIVES

Upon completion of this chapter, you will be able to:

- Lay out and construct block corners and jambs.
- Identify the special offset corner blocks and demonstrate their installations
- Demonstrate the proper alignment for block cut to length in a corner.

## Keywords

Block Jamb
Block Size
Checking the Range
Corner

Corner of the Lead
Lead
Nominal Size
Rack of the Lead

Tail of the Lead
Toothing

# Chapter Review Questions and Exercises

## SHORT ANSWER

1. To what potential hazards are those who carry and otherwise handle or are in close proximity to metal horizontal joint wire reinforcement exposed?

   _____
   _____
   _____

2. What personal protection equipment must one wear when operating masonry saws to cut block?

   _____
   _____
   _____

3. Describe the test for determining when mortar joints should be tooled.

   _____
   _____
   _____
   _____

4. What is the standard height for each course of block when modular masonry construction is intended?

5. Where it is necessary to increase or decrease course spacing, bed joint widths should be no less than _____ inch(es) and no more than _____ inch(es).

6. What tool is used to align CMU leads level, plumb, and straight?

7. What is held along the length of a wall to "check the range" between leads at opposite ends of a wall?

## COMPLETION

8. Block leads are typically built at opposite _____ of walls.

9. A _____ is a type of lead consisting of two walls adjoining at a 90° angle.

10. A block _____ is a type of lead typically found alongside a door or window opening.

11. Three block on one side and two on the adjoining side of a corner's first course permits building the corner a maximum of _____ courses.

12. Block are laid from the _____ of the lead toward the _____ of the lead.

13. A _____ line is used to align a lead straight with the opposite end of the wall.

14. Aligning the bottom edge of each block with the top edge of the block below it is known as _____ the block.

15. The three alignment operations as they are performed when building a lead are _____, _____, and _____.

16. The _____ of the lead is observed by holding the edge of the level diagonally with the top edge of each block at the tail end of each course.

17. The ends of 10-inch and 12-inch offset corner blocks are reduced to a width of _____ inch(es) to permit building walls having half-lap, running bond patterns.

18. A procedure known as _____ permits continuing a lead above the courses normally limited by the racking of the tail end.

19. Eliminating the construction of leads is sometimes accomplished by temporarily erecting metal or wood _____ _____, to which a mason's line is attached for laying block to the line.

20. Assume that you are a masonry contractor and upon arriving at a job site where you are to build CMU foundation walls, you discover that the concrete footings on which the corners are to be built are as much as 1 inch out of level. What can be done as each corner is constructed to ensure that the top of the last course of each is level with the tops of all others while complying with masonry standards?

    _____

    _____

    _____

    _____

21. Why should a mason use a masonry saw to rip the height of a masonry block, decreasing its height, rather than using a chisel and hammer?

    _____

    _____

    _____

    _____

22. Assuming it to be an 8-inch CMU lead having ⅜-inch wide head joints, give the lengths in inches of each of the courses of the lead in the following illustration.

#  Chapter 12 Estimating Masonry Costs

## OBJECTIVES

Upon completion of this chapter, you will be able to:

- Estimate quantities of brick, block, masonry cement, sand, and reinforcement.
- Estimate the amount of concrete needed for a concrete perimeter footing.
- Estimate the amount of materials needed for a concrete slab.
- Estimate the labor costs for given masonry projects.

## Keywords

Bid Price
Bidder
Cost-Estimate
Gable
Labor Constant

## Chapter Review Questions and Exercises

### SHORT ANSWER

1. For purposes of estimating materials, how many standard size brick are there per square foot? How many oversize brick per square foot? How many economy brick per square foot?

   _____

   _____

   _____

2. How much cement block is required for each square foot of wall area?

   _____

   _____

   _____

3. When mixed as recommended with other ingredients, how much brick do conservative estimates claim can be laid with one, 1-cubic-foot bag of masonry cement? How much block?

   _____

   _____

   _____

4. What is the minimum width for concrete trench footings supporting typical residential foundation walls? What is the minimum thickness for such concrete trench footings?

_____

_____

_____

5. What is the typical thickness of concrete slabs intended as residential pedestrian traffic such as those used for sidewalks and patios?

_____

_____

_____

## COMPLETION

6. The building code requires a minimum of one tie per _____ square feet of wall area.

7. One common nail capable of penetrating the wall stud a minimum of _____ inch(es) is required for anchoring each steel veneer tie.

8. What do contractors estimating the costs of larger projects rely upon for calculating masonry costs?

_____

_____

_____

9. The dollar amount of the proposal for supplying materials or labor for a specific job is called a _____ _____.

10. A triangular-shaped area below the sloping roofline of a building is called a _____.

11. One cubic yard of sand is estimated to mix _____ bags of masonry cement.

12. Concrete *quantities* are calculated in English unit of volume measurements expressed as "_____" of concrete.

13. One cubic yard is equivalent to _____ cubic feet.

14. The amount of welded wire fabric required to reinforce concrete is equivalent to the _____ of the slab.

15. A typical masonry wall requires _____ linear feet of horizontal wire joint reinforcement per 100 square feet of wall area.

16. The amount of labor required to perform a specific amount of work is known as the _____ _____.

17. List four factors influencing the cost of the labor required to complete a job.

_____

_____

_____

_____

18. The method in which the owner is obligated to pay the contractor for furnishing the materials and labor at a mutually established and accepted rate or price is known as "_____ and _____."

19. Assume that you are a masonry contractor currently obligated by many contracts. A homeowner asks you to give a price for providing materials and labor to construct block walls for a garage he intends to build. Knowing that you do not need to commit to any more work at the present time, you inflate the bid price considerably, realizing that the homeowner will most likely refuse the offer. You are willing to work it into your busy schedule if the homeowner does accept it. What can be the results of giving inflated bid pricing?

_____

_____

_____

20. Referring to the illustration below, estimate the number of standard-size brick needed to build this wall. The gable area is to be brick also. The wall includes one window, 3 × 5 feet, and one door, 3 feet × 6 feet 8 inches. Show all work.

_____

_____

_____

21. A wall measures 32 feet long and 12 feet high. How many of the following masonry units would be needed to build the wall?

    standard-size brick _____

    oversize brick _____

    CMUs _____

22. Calculate the number of cubic-foot bags of masonry cement needed to mix mortar for building a wall requiring:

    A. 19,500 standard size brick _____

    B. 2080 CMUs _____

23. A contractor needs 184 1-cubic-foot bags of masonry cement. How many cubic yards of sand are needed for mixing the entire 184 bags?

24. How many veneer wall ties are required for anchoring brick veneer to a backup wall measuring 24 feet long and 8 feet tall?

25. How many cubic yards of concrete are needed to place continuous concrete footings as illustrated below? It is assumed that the concrete footings are to have a thickness of 8 inches and a width of 2 feet.

26. How many cubic yards of concrete are needed to place a concrete slab measuring 24 × 32 feet, assuming the concrete is to be 4 inches thick?

# Chapter 13 Residential Foundations

## OBJECTIVES

Upon completion of this chapter, you will be able to:

- Identify factors considered for the design of concrete footings.
- Describe methods for forming concrete footings.
- List design elements for foundation walls built with concrete masonry units.
- Lay out and build a foundation wall.

## Keywords

6-6-10-10 Wire Reinforcement
Anchor Bolt
Anchor Strap
Areaway
Backfilling
Bar Reinforcement
Batter Boards
Brick Shelf
Compressive Strength
Crawl Space
Damp Proofing
Egress
Footings
Formed Footings
Foundation
Frost Line
Grout
Grouting
Parging
Potential Expansive Soils
Slump
Slump Cone
Slump Test
Trench Footings
Waterproofing

## Chapter Review Questions and Exercises

### SHORT ANSWER

1. What must be located, identified, and in some cases moved by the utility companies before excavation begins?

   _____
   _____
   _____

2. What personal protection equipment is required of those handling or exposed to reinforcement wire and steel bar reinforcement?

   _____
   _____
   _____

3. What precautions must those handling steel reinforcement take to avoid electrical shocks or fatal electrocution from overhead power lines?

4. What personal protection equipment must those who are exposed to airborne cement dusts wear?

5. With whose safety and environmental regulations and instructions must those exposed to foundation coatings comply?

6. What is required of anyone before operating backhoes and other mechanical equipment?

7. Because wet concrete can irritate and damage both skin and eyes, what personal protective equipment must be worn to prevent skin contact?

8. What must one do before entering ground excavations and trenches?

9. To what potential hazards are those lifting lintels exposed?

10. Why should the insides of foundation walls below grade be laterally braced until after supported structural framing and backfilling are complete?

11. Why may it be necessary for foundation walls above grade to be temporarily supported on both sides?

12. What precautionary measures should be taken to protect others, keeping them away from open excavations and projects under construction?

13. Complying with governing standards and building codes, who is responsible for the design of concrete footings?

14. In regions experiencing freezing temperatures, what is the minimum depth to which the bottoms of concrete trench footings must extend?

15. What type of welded wire reinforcement is placed in concrete slab foundations? Explain the design of this type of welded wire reinforcement.

16. What regulations govern the size of CMUs required for job-specific foundation walls?

17. What type or types of masonry cement are generally specified for mortars used to build CMU foundation walls?

18. What is the maximum spacing for threaded anchor bolts or sheet-metal straps intended to secure wood sill plates?

19. What is the maximum distance permitted from each end of a plate section for having an anchor bolt or plate strap?

20. What is the minimum diameter required for anchor bolts? What is the minimum depth for which they must extend below the top of a CMU foundation wall?

21. What is required for waterproofing the exterior sides of foundation walls?

22. What is the minimum ventilation requirement for ventilating a crawl space?

23. What is the recommended slump measurement for concrete intended for footings?

24. Depending on building code regulations and soil classifications, what may be placed in the open cells of CMUs to strengthen walls?

25. What type of container is used to form a sample of freshly mixed concrete to test the consistency or amount of water in the concrete?

26. What is used to cover fresh concrete for the first 24 hours, preventing excessive water evaporation from the concrete?

27. What is placed against the insides of foundation walls to help prevent walls from collapsing when backfilling?

_____

_____

_____

## COMPLETION

28. A _____ is that part of a building below the first floor framing.

29. Concrete _____ support the foundation walls and transfer the structure's weight to the soil.

30. Soil tests are conducted to determine the _____-_____ capacity of the soil.

31. Factors determining the size of a footing are: _____

_____

_____

32. Residential concrete footings are usually a minimum of _____ inches thick.

33. Soils that increase significantly in volume when wet are called _____ _____ soils.

34. Factors influencing the required compressive strength of concrete footings are:

_____

_____

_____

35. Steel rods embedded in concrete known as _____ _____ serve to strengthen the concrete.

36. The bottom of a concrete footing should be at least _____ inches below finished grade but also below the regional _____ line.

37. The _____ line for a given region is the greatest depth to which ground may be expected to freeze.

38. Concrete footing steps should be in _____-inch increments to permit full block coursings.

39. Factors governing the required block size for foundation walls are:

_____

_____

_____

_____

40. Mortar strength for block foundation walls should meet type _____ or type _____ requirements.

41. A mixture of cement and pea-size aggregates mixed with water to produce a consistency for pouring into blocks' open cells is known as _____.

42. Embedded in fully mortared block cells, _____ _____ or _____ _____ are placed for securing the wood framing to the foundation.

43. Damp-proofing the exterior walls of CMU foundations, offering minimum protection from water absorption, includes two procedures:

_____

_____

44. A plastic, _____-inch slotted drain line is a popular method for collecting water along a foundation's exterior walls.

45. Two materials above the drain line intended to prevent soil blockage are _____ and an approved _____ _____.

46. The foundations below houses having no basements are known as _____ spaces.

47. Measuring 8 × 16 inches, _____ _____ provide ventilation for enclosed foundation areas without basements.

48. A foundation _____ refers to the required exit from a basement to the outside.

49. A(n) _____ is a walled area below grade permitting ventilation, light, or an exit route.

50. Layout lines or strings identifying the locations for foundation walls are secured to _____ boards.

51. A ground excavation into which concrete is directly placed is known as a _____ footing.

52. Two methods of forming concrete footings are:

_____

_____

53. Steel bar reinforcement embedded horizontally in concrete footings should be placed nearer the _____ of the concrete to enable it to resist forces tending to crack the concrete.

54. Concrete should be covered for the first _____ hours after placing to prevent excessive water evaporation.

55. Designing foundation wall lengths in _____-inch increments eliminates cutting block to accommodate wall lengths.

56. Foundation walls supporting standard-size brick, anchored brick veneer walls require a _____-inch wide brick shelf for supporting the brick walls.

57. Veneer _____ _____ are embedded in foundation wall bed joints to anchor the brick veneer walls.

58. A concrete masonry _____ is placed across the top of an opening to support the wall above it.

59. It is recommended that concrete masonry lintels have a minimum bearing of _____ inches on each side of the opening.

60. For lintels having a groove or square slot along the center of the entire length of one side, the side having the square slot should be positioned in the wall as the _____ side of the lintel to maximize the embedded steel bar reinforcement's strengthening.

61. Solid, _____ block may be specified for the top course of a block foundation.

62. Two recommendations before backfilling a block foundation to help prevent the foundation walls from collapsing due to the pressures exerted by back-filled soils are:

_____

_____

63. Assume that you are a masonry contractor and that several children reside in the neighborhood where you are building a block foundation. It is obvious that both the children and adults are curious about the construction site, coming dangerously close to building materials and open excavations intended for below-ground foundation walls. What should you do to protect the unauthorized persons from open excavations and building materials?

_____

_____

_____

_____

64. Give the recommended minimum requirements for the concrete trench footing intended to support a residential CMU foundation wall and its placement in relationship to the finished grade level.

A. _____ -inch minimum and below _____ _____

B. _____ -inch minimum

C. _____ -inch minimum

# Chapter 14 Constructing Water-Resistant Brick Veneer Walls

## OBJECTIVES

Upon completion of this chapter, you will be able to:

- Identify the sources of water behind exterior masonry walls.
- Identify means of minimizing water migration through masonry walls.
- Define the terms *flashing* and *weeps*.
- Describe procedures for ensuring performance of flashing and weeps.
- List different materials used as flashing.
- Describe the proper procedures for installing flashing.
- Discuss the types of water repellants and their recommended applications.

## Keywords

| | | |
|---|---|---|
| Air Infiltration Barrier | Head Flashing | Sill Flashing |
| Air Space | Initial Rate of Absorption | Suction Rate |
| Backer Wall | Lime Run | Water Repellants |
| Base Flashing | Mortar Bridging | Weep Holes |
| Drainage Type Wall System | Mortar Collection Systems | Weep Vents |
| Efflorescence | Mortar Protrusions | White Scum |
| Film-Formers | New Building Bloom | Wick Ropes |
| Flashing | Penetrants | |

## Chapter Review Questions and Exercises

### SHORT ANSWER

1. To protect the hands from injury, what must be worn when handling metal flashings?
   _____
   _____
   _____

2. What must be worn when one is exposed to sealants in well-ventilated areas?
   _____
   _____
   _____

3. What additional personal protection equipment must be worn when exposed to fumes from sealants where adequate ventilation is not available?

4. Whose recommendations should be read and followed before applying any sealant to brick walls?

5. Whose regulations and standards must one comply with when using chemicals such as sealers that may be dangerous to individuals and the environment?

6. What is the initial rate of absorption (IRA) for brick at the time of laying that should not be exceeded?

7. What do most building codes and engineering specifications require for mortar joints?

8. What does the building code require as the minimum width for the air space between anchored brick veneer and the back-up wall?

9. What is the minimum lap for sections of flashing and how should they be sealed?

10. Where are the three locations where through-wall flashing is required?

11. What is the minimum diameter for open weep holes?

12. What is the recommended spacing between open weep holes and weep vents? Between wick ropes?

13. How many inches should flashing extend vertically above the fold where it is bedded in mortar?

14. How many inches should flashing extend beyond each end of steel lintels supporting brick above openings?

15. Water repellants should not be used to remedy water penetration on brick walls until what actions have been made?

16. List three damaging results to a wall system as a result of water migration.

## COMPLETION

17. It is reported that water migration through a brick wall occurs largely at separation cracks between the _____ and the _____ rather than through the brick themselves.

18. The _____ _____ of _____ is a test to determine the moisture content of a brick at the time of bedding in mortar.

19. Applying _____ head joints and avoiding _____ furrowed bed joints helps to reduce water migration through a brick wall.

20. Mortar _____ is observed within the air cavity when excessive amounts of mortar in which brick are bedded obstructs the air cavity.

21. Water impermeable materials known as _____ are installed in masonry walls to control water migration.

22. Intended to prevent water from draining into the open cells of a block foundation wall or possibly damaging wall framing, _____ flashing is installed just above finished grade.

23. Openings in head joints permitting water drainage from the air cavity between the brick facade and the backup wall to the exterior of the brick wall are called _____ holes.

24. List three problems that may be experienced with having open head joints intended as weep holes.

_____

_____

_____

25. Cotton _____ _____ are sometimes used to route water from the air cavity to the face of the wall.

26. Installing _____ _____ within head joints above base flashing permits water drainage from the air cavity and promotes air circulation within the air cavity while blocking the passage of insects into the wall's air cavity.

27. Installed below window frames and door sills, _____ flashing diverts migrating water from the air cavity and to the exterior of the brick wall.

28. _____ flashing helps control corrosion and rusting of the steel lintels above door and window frames and also prevents water from damaging window and door frameworks.

29. Products designed to prevent mortar droppings from blocking weep holes or weep vents are called _____ _____ _____.

30. A white, powdery stain on the face of masonry walls as a result of water-soluble salts appearing after water evaporates is called _____.

31. List three sources for the water-soluble salts found on the surface of masonry walls.

_____

_____

_____

32. The presence of increased water during construction can result in white-colored stains on the exterior faces of walls called new _____ _____.

33. Water repellants called _____-_____ coat the surface of a wall but do not penetrate the wall.

34. Water repellants called _____ are absorbed into masonry walls.

35. Identify the following characteristics as those of *film-formers, penetrants,* or perhaps both.

    _____ : permit evaporation of moisture from within the masonry wall

    _____ : prevent water from entering hairline cracks

    _____ : good resistance to ultraviolet degradation

36. List four details requiring attention for eliminating water migration through masonry walls.

_____

_____

_____

_____

37. Assume that you are a masonry contractor and you are aware of another masonry contractor who is facing legal action from the home's builder because the homeowner is experiencing water intrusion in the new home during heavy rains, so much that the interior wall board on exterior walls is saturated

with water. Reports conclude that inferior masonry construction is the probable cause for the water intrusion. What can you do to prevent such litigation, ensuring the brick facades that you and your employees build comply with all building codes and recommended industry standards?

38. The thickness of flashing materials is expressed in metric units called millimeters. Given that ¹⁄₁₆ inch is equivalent to 1.588 millimeters, and knowing that typical flashing materials measure between 0.5 millimeters and 1 millimeter, which of the three fractions best describes the thickness availability for a flashing material expressed in the English unit fractions of an inch?

   A. ¹⁄₆₄ inch

   B. ¹⁄₃₂ inch

   C. ⅛ inch

39. Knowing that 1.588 millimeters is equal to ¹⁄₁₆ inch, how many millimeters are equivalent to 1 inch?

40. Knowing that it is recommended to place an open weep between the brick course above base flashing at 24-inch intervals, how many standard-size brick, modular-size brick, or queen-size brick are between adjacent open weeps?

41. For a house measuring 48 feet long and 26 feet wide, which of the following estimations is correct for the quantity of wick ropes needed above the base flashing if wick ropes are spaced 16 inches apart?

   A. 75

   B. 112

   C. 175

42. Which of the following should be the minimum width for through-wall base flashing used on a 4-inch anchored brick veneer wall having a 1-inch air space between the brick wall and the back-up supporting wall?

   A. 8 inches

   B. 10 inches

   C. 14 inches

# Chapter 15 Anchored Brick Veneer Walls

**OBJECTIVES**

Upon completion of this chapter, you will be able to:

- Identify factors influencing the spacing of brick veneer from the back-up wall.
- Identify and describe different methods for anchoring brick veneer to supporting back-up walls.
- Demonstrate the use of masonry spacing scales for proper brick course spacing.
- Demonstrate base, sill, and head flashing installations.
- Set steel lintels and lay brick above openings.
- Identify areas where differential movements are likely to occur and describe brick veneer construction details necessary for permitting differential movements of building materials.
- Construct brick veneer walls.
- Construct brick rowlock sills.

## Keywords

| | | |
|---|---|---|
| Adjustable Anchor Assemblies | Differential Movements | Load-Bearing Wall |
| Air Infiltration Barrier | Drip | Masonry Sills |
| Anchored Brick Veneer | Eaves | Non-Bearing Wall |
| Arching Action | Facia | Rake Board |
| Brick Binding | Frieze Board | Soffit Board |
| Brick Mold | Gable | Steel Angle Lintel |
| Brick Shelf | Head | Veneer Wire Anchoring System |
| Corrugated Steel Ties | Jamb | Wind Loads |
| Dead Loads | Live Loads | |

## Chapter Review Questions and Exercises

### SHORT ANSWER

1. What precautions should be taken before using caulking products?

    _____
    _____
    _____

2. Who should one receive training from before using ladders?

3. From where should the presence of metal or electrical conductive ladders be restricted?

4. What potential hazard is one exposed to when hands and arms are in close proximity to anchored veneer wall ties?

5. What is the minimum distance to maintain between steel wire joint reinforcement and overhead power lines?

6. Other than its own weight, what loads are anchored brick veneer walls designed to support?

7. What is the required minimum width for the air space between anchored brick veneer walls and wood-framed back-up walls?

8. What is the required minimum width for the air space between anchored brick veneer walls and steel-stud back-up walls?

9. What is the maximum distance for which anchored brick veneer walls are permitted to extend beyond the front edge of the brick shelf?

**CHAPTER 15** Anchored Brick Veneer Walls  **67**

10. What is attached to the exterior sides of exterior wall framing to minimize air infiltration?

11. What depth are nails securing veneer wall ties to penetrate wood wall stud framing?

12. What is the maximum vertical spacing, spacing the length of a brick course, for veneer wall ties?

13. What is the typical horizontal spacing, or number of courses, between veneer wall ties when using standard-size brick? When using oversize brick?

14. What percentage of the bed depth, or veneer thickness, should veneer ties penetrate?

15. What is the minimum distance from the face of an anchored brick veneer wall for veneer ties to be bedded in mortar?

16. What wall anchorage system is used to anchor brick veneer to steel framing?

17. What should be the minimum slope for a brick rowlock window sill?

18. What should be the minimum extension for a masonry sill beyond the face of the wall?

19. For wood-framed structures, what width should the caulked joint between a masonry sill and the bottom edge of an adjoining window or door frame be to accommodate differential movement of building materials?

20. What should be the minimum thickness of steel used as masonry lintels?

21. What width is the horizontal leg of steel angle lintels to be for supporting standard-size brick above openings?

22. What factors determine the width of the vertical leg of steel angle lintels?

23. For spans no greater than 4 feet, what should be the minimum bearing for the ends of the angle lintels to rest on the supporting walls? For spans exceeding 4 feet?

24. What is a pointed tool suspended from a string line that is used to establish a plumb line called?

25. What tools are intended to mark course spacing, possibly preventing cutting brick below sills and at the heads of openings?

26. Depending upon the span of the opening below it, what should be temporarily placed below the center of a steel angle lintel to prevent it from deflecting or sagging?

## COMPLETION

27. A single-wythe brick wall, requiring vertical support and anchored to either a load-bearing or non-bearing wall, having an air space between the brick wall and the back-up wall is called _____ _____ _____.

28. A _____-_____ wall is one designed to support its own weight and the weight and forces placed upon it.

29. Loads imposed on a structure including occupants, furnishings, rain, and snow are examples of _____ loads.

30. Structural framing, wallboard, floor and roof systems, and permanent attachments such as air conditioning systems are examples of _____ loads.

31. Anchored brick veneer walls are classified as _____-_____ walls since they support no loads or weight other than their own.

32. Before bedding the first course in mortar, the brick are _____ _____ to establish uniform brick spacing having properly sized head joints.

33. Relying on the masons' brick coursing _____ ensures appropriately sized, uniform bed joints while allowing constructing walls to desired heights.

34. Moisture-related problems within wall systems are controlled by installing:
    a. _____, _____, & _____ flashings
    b. _____ holes/vents or _____ ropes
    c. _____ deflection/collection systems
    d. _____ _____ barrier

35. A corrosion-resistant, veneer wire anchoring system is designed to anchor brick veneer to both wood-framed and metal-stud framed back-up walls, having air cavities up to _____ inches wide where the supporting brick shelf design permits.

36. Brick is typically bedded in the _____ position, face side up, immediately below window frames and door frames to cap the top of brick coursing below the sill frames.

37. The vertical, upright side of a door or window frame is called the _____.

38. At window and door jambs, a _____-inch spacing between the ends of brick and frames accommodates differential movements of materials and permits a caulked, water-resistant barrier between the brick and the frames.

39. The top of a masonry opening is called the _____ of the opening.

40. Supported by brick walls on both sides of an opening, a steel _____ _____ is placed horizontally across an opening to support brickwork continuing above the opening.

41. A condition known as _____ action permits part of the brick facade above an opening to become self-supporting once the mortar cures.

42. Two concerns presented with the procedure of toothing at corners are:
_____
_____

43. The triangular areas below rooflines are known as _____.

44. As a mason, what should you do upon discovering that the air infiltration barrier or house wrap is torn in an area where you are supposed to continue building a brick wall?

45. Assume that you are a masonry contractor and upon inspecting the plumb alignment of the exterior wall house framing to which you are to attach anchored brick veneer walls, you discover that the house framing is as much as 1 inch misaligned from being plumb. To prevent conflicts with the homebuilder, homeowner, or possibly the building inspector, what course of action should you take before starting the brick walls?

46. As a masonry contractor, what should you do if the nails provided to you for attaching veneer wall ties to wood studs are not the adequate length for complying with building codes?

47. Assume that a masonry contractor fails to observe recommended clearances between brick sills and the window manufacturer's window sill. What litigation may arise from this type of non-compliance?

48. The maximum allowable projection for the face of the brick beyond the supporting brick shelf is the lesser of ½ the brick's height or ⅓ the brick's width. Assuming that the height of a brick measures 2¼ inches and its width measures 3⅜ inches, what is the maximum projection for the face of the brick beyond the supporting brick shelf?

49. For a 4-inch anchored brick veneer wall weighing 30 pounds per square foot, what weight does the steel angle lintel support assuming that arching action occurs and as a result the only weight supported by the steel angle lintel is an area represented by an equilateral right triangle whose area is calculated using the formula Area (A) = ½ Base (B) × Height (H)?

_____

_____

_____

_____

50. Assuming arching action occurs, the approximate weight supported by the steel angle lintel is

_____ pounds.

51. Provide the missing information as recommended by industry standards or code requirements. Anchored brick veneer wall tie details:

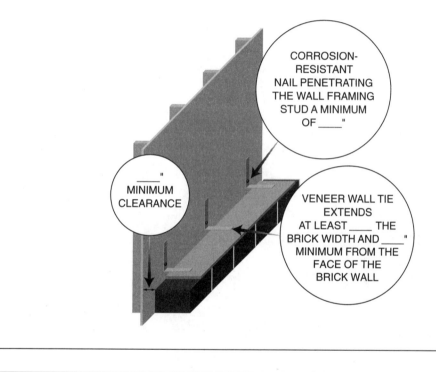

_____

_____

_____

_____

# Chapter 16 Composite and Cavity Walls

**OBJECTIVES**

Upon completion of this chapter, you will be able to:

- Compare similarities and differences between cavity walls and composite walls.
- Identify the types of thermal insulation used in cavity walls and composite walls.
- Explain how a rain screen wall is designed to reduce water migration through its exterior wythe.
- Identify the types of joint wire reinforcements and their applications.
- Explain how brick veneer walls with masonry backing are different from cavity walls or composite walls.
- Explain the differences between control joints and expansion joints and where each is used.
- List the procedures for constructing both cavity walls and composite walls.
- Explain how grout is used to reinforce cavity walls and composite walls.

## Keywords

| | | |
|---|---|---|
| Adjustable Assemblies | Differential Movement | Multi-Wythe Grouted Masonry Wall |
| Bearing Plates | Drainage Wall Systems | Parapet Wall |
| Bond Beam | Drip | Rain Screen Wall |
| Cap | Expansion Joint | Rigid Insulation |
| Capital | Fire Wall | Shelf Angles |
| Cavity Wall | Granular Fill Insulations | Soft Joints |
| Cleanouts | High-Lift Grouting | Through-Wall Flashing |
| Composite Masonry Wall | Ladder-Type Wire Reinforcement | Truss-Type Wire Reinforcement |
| Control Joint | Load-Bearing Wall | Wythe |
| Coping | Low-Lift Grouting | |

## Chapter Review Questions and Exercises

### SHORT ANSWER

1. What precautions must one take when using adhesives such as those intended to bond sections of flashing materials?

   _____
   _____
   _____

2. What personal protection must those exposed to granular fill insulations wear to avoid breathing airborne particles and to keep airborne dust from injuring the eyes?

3. What precautions must one take while holding on to large pieces of insulation foam board if high winds prevail or are expected?

4. What precautions must one take when handling wire joint reinforcement to prevent electric shock or fatal electrocution from overhead power lines?

5. What precautions must one take when handling wire joint reinforcement to prevent impaling other workers with its pointed ends?

6. Who should one consult for recommended grouting procedures and precautions before grouting walls?

7. By whom should one intending to perform grouting operations be trained before grouting and using mechanical grouting equipment?

8. Walls grouted within what time period require bracing by a competent person to prevent possible "blow-out," the unintentional separation of the masonry units caused by forces exerted by the wet grout?

9. What are the minimum and maximum widths of the air space for walls classified as cavity walls?

10. How many inches should flashing extend vertically above its base?

11. How many inches should adjoining sections of flashing be lapped?

12. Describe how the ends of flashings above and below openings should be folded. Why is this recommended?

13. What is the recommended spacing for open weep holes and weep vents? For wick ropes?

14. According to industry recommendations, how many inches should the ends of wire joint reinforcement be lapped?

15. Describe how a typical expansion joint is filled.

16. What percentage or fractional part of a brick's bed depth must shelf angles support?

17. What masonry spacing scales ensure aligning the bed joints of two dissimilar materials at 16-inch height intervals?

18. What height interval is horizontal wire joint reinforcement typically placed in cavity walls or composite walls?

19. What procedure is recommended for the cavity or collar joint below base flashing found in cavity walls or composite walls?

20. With what should the joints between sections of pre-cast concrete and stone copings be filled?

21. In what direction is water diverted by the sloping top of wall coping, toward the roof or toward the face of the wall? Why is this runoff direction better?

22. What are two means of mechanically consolidating grout?

## COMPLETION

23. A _____ wall consists of two adjoining masonry wythes reinforced with horizontal wire reinforcement to perform as a single wall in resisting forces and loads.

24. The adjoining wythes of _____ walls are separated by an air space no less than 2 inches wide.

25. A _____ wall is a fire-resistant wall built from the foundation and extending above the roof to restrict the spread of fire.

26. Designed to divert wind-driven rains migrating through the outer wythe to the outside, cavity walls are classified as _____ wall systems.

27. A system of _____ and _____ are designed to channel water from the air cavity to the exterior of the outer wythe.

28. Two types of cavity wall insulation are _____ _____ and _____.

29. A _____ _____ wall is a classification of a masonry cavity wall containing weep vents at the bottom and protected air vents near the top, permitting equal air pressures of the outside air and that within the air cavity.

30. The three types of composite and cavity wall horizontal wire joint reinforcement are:

    _____

    _____

    _____

31. Because it may restrain differential movements between wythes, _____-type wire joint reinforcement is not recommended for cavity walls.

32. Completing the inner wythe before beginning the outer wythe to reduce the time required for enclosing a building is a major advantage for using _____ _____ types of reinforcements.

33. Because clay and shale brick expand over time, _____ joints divide longer brick masonry walls into shorter sections.

34. _____ joints are placed along concrete block masonry walls to control the location of cracks resulting in the shrinkage of CMU walls.

35. Welded or riveted to steel I-beams, _____ _____ support and transfer the weight of masonry walls to the structure's steel framework.

36. Referred to as _____ joints, these horizontally positioned expansion joints minimize wall cracks by permitting both the expansion of brick masonry and the deflection of steel shelf angles.

37. Using the _____ course spacing scale system ensures aligning the bed joints of two dissimilar masonry materials level with one another at 16-inch intervals, permitting the installation of horizontal wire reinforcement across the two adjoining wythes.

38. Flashing is placed above steel lintels and shelf angles to prevent water from _____ the steel, decreasing the lintel's intended load capacity.

39. A _____ wall is that part of an exterior wall extending above the roofline.

40. The cavity and top of double-wythe walls are covered with stone, pre-cast concrete, or metal materials known as _____.

41. A _____ is a covering for part of or an entire wythe that is terminated below the top of a wall.

42. A _____ is a cutout in the underside of projecting stone or concrete coping that prevents water from traveling back to and running down the face of the wall.

43. Assume that you are a masonry contractor. The horizontal wire joint reinforcement being supplied for the exterior masonry cavity walls that you are responsible for building appears to be uncoated steel rather than "hot-dipped" galvanized. What should you do, use the supplied wire reinforcement anyway or bring the concern to the attention of the builder?

    What long-term effects could uncoated reinforcement have on the exterior cavity walls?

    How could this effect your reputation as a masonry contractor?

    _____

    _____

    _____

44. As a masonry contractor, you observe that job site workers are handling 10-foot sections of wire joint reinforcement stockpiled below low, overhead power lines. What should you do immediately?

_____

_____

_____

45. Modular wall construction requires 6 courses of standard-size brick to be equivalent to a height of 16 inches. What is the height of one course of brick? _____

   Is this measurement inscribed in inches and fractions on the mason's rule? _____

46. Modular wall construction requires 5 courses of engineered-size brick to be equivalent to a height of 16 inches. What is the height of one course of brick? _____

   Is this measurement inscribed in inches and fractions on the mason's rule? _____

47. Modular wall construction requires 4 courses of economy-size brick to be equivalent to a height of 16 inches. What is the height of one course of brick? _____

   Is this measurement inscribed on the mason's rule? _____

# Chapter 17  Brick Paving

### OBJECTIVES

Upon completion of this chapter, you will be able to:

- Describe different types of paving brick.
- List the elements of a brick paving system.
- Compare similarities and differences between the types of bases for brick paving systems.
- Describe the procedures for installing a rigid brick paving system.

## Keywords

Basket Weave Pattern
Brick Density
Brick Pavers
Classification MX Paving Brick
Classification NX Paving Brick
Classification SX Paving Brick
Control Joint
Coping Brick
Expansion Joint
Field Pattern

Flashed Brick
Flexible Base
Grouting
Heavy Traffic
Herringbone Pattern
Light Traffic
Medium Traffic
Modular Paving Brick
Mortared Brick Paving
Mortarless Brick Paving

Pattern Bond
Repressed Paving Brick
Running Bond Pattern
Semi-Rigid Base
Skid Resistance
Slip Resistance
Spalling
Stacked Bond Pattern

## Chapter Review Questions and Exercises

### SHORT ANSWER

1. What possible buried items must be located and identified in and surrounding the area of construction before ground digging begins?
   _____
   _____
   _____

2. Whose regulations must one be in compliance with when using, storing, and disposing of hazardous chemicals?
   _____
   _____
   _____

3. What personal protection equipment must those who are exposed to chemical cleaners wear?

4. Based on resistance to weathering, what are the three classifications of paving brick?

5. Based on its intended use, what are the three traffic classifications of paving brick?

6. What are the surface dimensions of modular brick pavers?

7. What is the recommended maximum spacing of control joints between concrete slabs supporting mortared brick paving?

8. What type of masonry cement is recommended for mortared brick paving?

9. What power tool can be used to fill the mortar joints between brick pavers that have been bedded in mortar?

10. What manually operated equipment can be used to fill the mortar joints between brick pavers that have been bedded in mortar?

## COMPLETION

11. Brick pavers used for walkways and patios are referred to as either paving brick or brick _____.

12. Brick classified as _____ should be used on exterior paving wherever freezing may occur.

13. Depending on intended use, paving brick are classified as _____, _____, or _____ traffic applications.

14. Paving brick intended for residential pedestrian use must meet _____ traffic application specifications.

15. The approximate thickness of most pedestrian pavers is either _____ inches or _____ inches.

16. Intended for high-traffic areas, _____ paving brick are first extruded and then compressed to increase density.

17. The measure of pedestrian traction on a wet surface is called _____ resistance.

18. List two brick manufacturing processes used to improve traction on a wet surface.
    _____
    _____

19. List the three components of a properly designed rigid-base system.
    _____
    _____
    _____

20. A leveling sand bed and a gravel sub-base forms a _____ base.

21. Paving brick measuring _____ inches wide and _____ inches long permit pattern uniformity for mortarless paving of the basket weave pattern.

22. List three pattern bonds for brick paving.
    _____
    _____
    _____

23. A _____ texture promotes a good bond between the mortar and the concrete base.

24. A _____ joint permits masonry to expand and contract at a formed or sawed separation.

25. The brick pattern between the borders is called the _____ pattern.

26. List three techniques when laying the pavers to increase bond strength.
    _____
    _____
    _____

27. Give an advantage of waiting at least 24 hours to fill mortar joints between brick bedded in mortar.
    _____

28. List two reasons for overfilling the mortar joints when grouting.

29. Type _____ masonry cement is recommended for brick paving.

30. Give one advantage for tooling the mortar joints. _____

31. Assume that you are a masonry contractor and a homeowner wants you to lay mortared brick paving on an existing concrete patio adjoining their home. The concrete slab appears to be structurally sound, exhibiting no cracks or surface deterioration. However, soils below the concrete slab along the outside of the foundation walls have settled, allowing its top surface to slope toward the house rather than away from it. Rain water currently drains toward the home's foundation walls rather than away from them. The homeowner is hoping that you can remedy this problem by bedding the brick nearer the house in much thicker mortar, thereby sloping the paved brick surface away from the house. You conclude that this requires bedding the brick pavers in 1½-inch-thick bed joints, much more than is recommended. What would be your response to the homeowner?

32. A homeowner questions you, a masonry contractor, about your different labor rates, including labor costs per square foot, for laying different patterns. What valid reasons can you give for the rate differences?

33. Assume that you are a masonry contractor meeting with a homeowner to give a contract price for laying mortared brick paving on an existing concrete walkway. You notice that branches of shrubbery extend over the concrete walkway, inhibiting access along most of its edges. What suggestions do you have for the homeowner who is opposed to cutting and removing the branches?

34. Determine the quantity of modular brick pavers required to cover a walkway 36 feet long and 4 feet wide.

35. Determine the quantity of modular brick pavers required to cover a concrete patio whose dimensions are 50 × 30 feet.

# Chapter 18: Steps, Stoops, and Porches

**OBJECTIVES**

Upon completion of this chapter, you will be able to:

- Identify and define the components of brick steps.
- List the procedures for constructing brick steps.
- Identify factors to consider when planning the dimensions for brick steps.
- Describe methods of providing foundation support below brick steps.
- Describe methods for constructing porches and stoops.

## Keywords

Loose Fill
Porch
Rise
Riser
Slab Footing
Stoop
Tread
Tread Depth

## Chapter Review Questions and Exercises

### SHORT ANSWER

1. Who must one notify and what must be done before ground digging?

   _____
   _____
   _____

2. What is the recommended means for lifting and moving concrete masonry lintels?

   _____
   _____
   _____

3. What are the OSHA regulations that one is to comply with when working below grade in trenches created by foundation walls and soil embankments?

   _____
   _____
   _____

4. What precautionary measures should be taken to protect others, keeping them away from projects under construction?

## COMPLETION

5. The rise of a step should not be more than _____ inches.
6. The concrete footing below most steps should have a thickness of _____ inches and extend _____ inches beyond the exterior wall perimeter.
7. Treads should have a slope of _____ inches per foot toward their front edge to allow water runoff.
8. The thickness of a concrete slab intended as the top surface for residential porches and stoops is typically _____ inch(es).
9. The height between steps is called the _____.
10. One's feet are placed on the part of a step called the _____.
11. The "rule of _____" is a guideline for calculating tread depth.
12. Varying the width or thickness of the _____ between brick courses enables the mason to slightly alter the riser heights.
13. Hollow concrete masonry units placed below grade should be filled with _____ to prevent ground water accumulation within their open cells, leading to deterioration.
14. Ground soil removed during construction and later replaced without compaction is known as _____.
15. Concrete masonry _____ are used to bridge loose fill at the foundation wall and support the weight of walls above them.
16. A _____ is a sheltered area at the entrance of a building.
17. A _____ is a platform at the entrance of a building.
18. A homeowner insists that you build a set of steps using a design that does not comply with building codes, having tread depths that are less than required and riser heights of more than 8 inches. The homeowners assure you that they will not hold you responsible for any accidents that may result from the faulty design. What should you do?

19. Assume that you are a masonry contractor. The builder of a new home has placed a concrete slab footing on loose backfill, on which he intends for you to build brick steps. What should you do?

20. How can one's reputation as a masonry contractor become distrusted over time by choosing to proceed with masonry construction that does not comply with building codes or industry standards? What then is the best course of action to take when it is apparent that the intended job design is substandard, failing to comply with building code requirements?

___

Identify the numbered items.

1. _____
2. _____
3. _____
4. _____ inches
5. _____
6. _____
7. _____
8. _____
9. _____ inches
10. _____
11. _____
12. _____

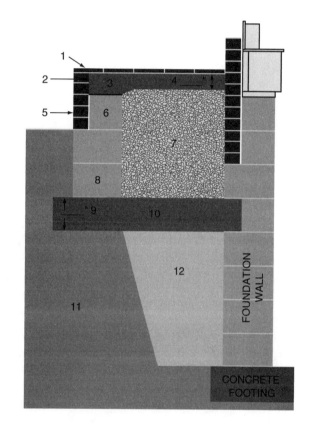

# Chapter 19: Piers, Columns, Pilasters, and Chases

## OBJECTIVES

Upon completion of this chapter, you will be able to:
- Define the terms pier, pilaster, chase, and column.
- Identify and give uses for masonry piers, pilasters, chases, and columns.
- Lay out and build a pier, pilaster, chase, and column.

## Keywords

Actual Size
Capital
Capping
Chase
Column
Compass Brick
Compressive Strength
Concentrated Load
Control Joint
Hollow Masonry Pier
Lateral Strength
Masonry Column
Nominal Size
Pier
Pilaster
Pilaster Block
Radial Pier
Reinforced Masonry Pier
Structural Pier

## Chapter Review Questions and Exercises

### SHORT ANSWER

1. Who must one notify and what must be done before ground digging as when placing footings to support masonry piers or columns?

   _____
   _____
   _____

2. What precautionary measures should be taken to protect others, keeping them away from open excavations and projects under construction?

   _____
   _____
   _____

3. With whose regulations must masonry piers or columns such as those containing mailboxes and near roadways be in compliance?

_____

_____

_____

4. What are the defining characteristics of a masonry column?

_____

_____

_____

## COMPLETION

5. A masonry _____ is a structural support below another structural element.

6. The structural weight supported by a pier is called a _____ load.

7. A _____ is a pillar of masonry projecting from a wall.

8. Brick having a curved face rather than a straight face and used for round piers are called _____ brick.

9. A _____ masonry pier should not be used as a structural support supporting structural loads.

10. A _____ masonry pier is strengthened with concrete and steel reinforcement.

11. To be considered a column, a masonry pier must be at least _____ times taller than its least lateral dimension.

12. A measure of the vertical weight or force from above that a wall can support is known as the _____ strength of the wall.

13. A measure of the horizontal force (force applied to the face of a wall) that a wall can resist is known as the _____ strength of the wall.

14. Usually made of stone or precast concrete, a _____ is a decorative element found at the top of a pier, pilaster, or column.

15. A load-bearing pier or pilaster should be designed by a licensed, structural _____.

16. All four corners of a pier are square if the lengths of the _____ measurements are equal.

17. For a pier whose length and width is less than that of the mason's level, height or scale (course spacing) should be checked at _____ corner(s).

18. The face of a pilaster forms a line that is typically aligned _____ with the wall line.

19. A _____ pier or column is one having no reinforcement.

20. A _____ pier or column is one strengthened with steel and concrete.

21. A _____ block is a special-shape CMU designed to stabilize a CMU wall and provide a control joint for the wall.

22. A _____ joint regulates the location of separation in a CMU wall caused by dimensional changes of the building materials.

23. A masonry _____ is a recess in a wall designed to provide space for electrical, plumbing, and heating systems.

24. Assume that you are a masonry contractor. You are to construct brick entry columns to the sides of the home's driveway entrance. You explain to the homeowner that governing laws require that local utility companies or their agents be notified prior to digging, a necessity for placing the concrete footings for supporting the columns. The homeowner assures you that there are no buried utilities, water lines, or sewer lines near the area to be excavated, encouraging you to proceed with the excavations for the two columns. What should you do?

25. Suppose a homeowner asks you, a masonry contractor, to build a brick column enclosing a mailbox at a public roadside. What should you have the homeowner document in writing to you before building the brick mailbox column?

26. Why is it a good idea for a masonry contractor to request from the homeowner printed copies of documents validating governing authority's approval for the location and design of such things as brick entry columns and mailbox posts?

27. It is a false assumption that the corners of a four-sided object must be square just because the front side length is equivalent to the backside length and the left side length is equivalent to the right side length. The pair of opposite sides can be parallel and therefore have equivalent lengths. Only when the lengths of the diagonal measurements are equivalent can all four corners be square, meaning that the adjacent sides are at 90° angles.

Perform the four-step procedure as explained in Chapter 19 for laying out and constructing piers and columns, making a layout on a concrete slab for a pier whose dimensions are 16 × 20 inches. Have the instructor check your layout when finished.

# Chapter 20 Appliance Chimneys

**OBJECTIVES**

Upon completion of this chapter, you will be able to:

- Identify the parts of brick masonry chimneys.
- Explain important regulations and codes governing the construction of masonry chimneys.
- Construct an appliance chimney.

## Keywords

Appliance Chimney
Chimney Base Flashing
Chimney Cap
Cleanout
Corbelling
Counter-Flashing
Cricket
Cross-Sectional Area
Fireblocking
Fireclay
Flue Lining
Thimble

## Chapter Review Questions and Exercises

### SHORT ANSWER

1. Before descending below grade level into ground excavations, what must be done where there is a potential for cave-ins and where excavations are 5 feet or more in depth?

2. To whom should the erection, dismantling, or accessing of scaffolds be limited?

3. What is the minimum spacing between scaffolds and overhead power lines?

4. At what height are scaffolds to be adequately braced or tied to restrain and prevent them from tipping?

5. What personal protection equipment must those who are exposed to the cutting of masonry materials with power saws have?

6. Although covering the tops of flues prevents debris from falling into chimney flues, why should the flues of chimneys never be blocked in ways to prevent them from venting gases and products of combustion when replacing chimney caps?

7. At what heights above lower levels are workers required to have 100% fall protection?

8. Why is it important to have a competent person inspect and clean the interior walls of flue linings routinely, especially those flues venting appliances that burn solid fuels, preferably no longer than 12 months between inspections?

9. What are the design requirements for a concrete footing intended to support an appliance chimney?

10. What type of masonry units are required for the walls of a masonry chimney?

11. What codes govern the mortar type requirement for masonry chimney walls?

CHAPTER 20 *Appliance Chimneys* 93

12. What should be the width of the air space or insulation separating the flue lining and the interior face of the chimney walls?

13. What determines the required minimum size of the flue liner for an appliance chimney?

14. In relationship to finished grade level and the location of a chimney's thimble, what are the requirements governing the location of a chimney cleanout door?

15. What should be used as bedding mortar between each flue liner section?

16. What is the minimum height that a chimney should extend above the highest point where it penetrates the roofline?

17. What is the required minimum width for an air space between the exterior side of chimney walls and nearby interior combustible building materials?

18. For chimneys located entirely on the outside of exterior walls constructed of combustible materials, what is the required minimum width for an air space between the outside of the chimney walls and the combustible materials?

19. When is a chimney cricket required for chimneys whose lengths are parallel to but not intersecting the roof ridgeline?

20. What power tools and their accessories are recommended for cutting round holes through the walls of a flue lining to accommodate the connecting thimble?

_____

_____

_____

## COMPLETION

21. _____ chimneys differ from fireplace chimneys as they are designed to vent appliances burning solid fuels, natural gas, propane gas, and fuel oil.

22. The bottom of the footing must be below the _____ line or a minimum of _____ inches below finished grade in areas not subject to freezing.

23. A single-wythe wall having a nominal thickness of _____ inches is permitted for chimneys.

24. A _____ _____ is a rectangular or round, hollow unit intended to contain the combustion products and protect the chimney walls.

25. Sections of flue linings are joined by using medium-duty _____ mortar.

26. The size of rectangular flue liners refers to the _____ dimensions while the size of round flue liners refers to the _____ diameter.

27. A fireclay _____ provides connection between the flue lining and the appliance connector pipe.

28. A chimney flue lining must begin at least _____ inches below the thimble.

29. A chimney _____, made of reinforced concrete, eliminates moisture penetration at the top of the chimney.

30. Allowing the concrete footing to cure for a minimum of _____ days before constructing a chimney permits the concrete to strengthen.

31. A _____-inch solid masonry wall, bonded into the walls of the chimney, is typically required between multiple flues in a chimney.

32. Align a thimble _____ with the inside of the lining, bonding the two with _____ cement.

33. Thimbles should be placed a minimum of _____ inches below ceilings.

34. A minimum of _____ inches of solid masonry is required enclosing the outer sides of thimbles passing through masonry walls.

35. Thimbles passing through walls of combustible materials such as wood framing must be enclosed in a minimum of _____ inches of self-supporting solid masonry.

36. Connecting only _____ appliance(s) to a single flue is permitted.

37. Placing a minimum _____-inch layer of noncombustible material to serve as fireblocking between chimneys and floor or ceiling framing is required.

38. Bedded at height intervals of _____ inches between mortar joints, chimney anchorage to structural framing must conform to building code requirements.

39. The last flue lining should extend above the finished brickwork to permit a _____-inch to _____-inch lining extension above the chimney cap.

40. Placing flashing on top of the brickwork and below the cap serves as a _____ break, isolating the concrete cap from the brickwork and providing for differential movements of brick and concrete.

41. A water-sealed _____-inch-thick layer of noncombustible, compressible material such as fiberglass insulation between the concrete cap and the flue liner permits the lining to expand without cracking.

42. Type _____ mortar is recommended as a bedding mortar for prefabricated concrete caps.

43. Installing a removable _____ cap prevents rain, leaves, birds, and rodents from entering the chimney.

44. Corrosion-resistant sheet metal known as _____-flashing is embedded in bed joints to limit water invasion between the roof and chimney walls.

45. Chimney counter-flashing overlaps the _____ flashing.

46. A _____ is a double-sloped roof behind the upper side of the chimney parallel with the ridgeline diverting water from behind the chimney toward the sides.

47. Assume that you are a masonry contractor. The homeowners are asking you to build a masonry appliance chimney as they have designed it. You point out to them that the design does not comply with building code requirements. The homeowners insist that they will take the responsibility for the chimney's noncompliant design should problems arise. What should you do?

_____

_____

_____

48. Assume that you are a masonry contractor. The masonry chimney that you are building is on the outside of a home and close to overhead power lines coming to the home. The supported scaffold you need to erect is out of reach but within 10 feet of the power lines. What should you do before erecting the scaffold to protect workers from possible electric shock or fatal electrocution should contact be made with a power line?

_____

_____

_____

49. Provide the missing information for chimney top details.

   a. _____

   b. _____ inches to _____ inches

   c. _____

   d. _____ foot minimum

   e. _____

   f. _____ inches

   g. _____

   h. _____

# Chapter 21 Masonry Fireplaces

## OBJECTIVES

Upon completion of this chapter, you will be able to:

- Identify the components of a wood-burning fireplace.
- Explain basic features of the four types of masonry fireplaces.
- Explain factors governing the performance of a fireplace.
- Explain building code requirements for a single-face masonry fireplace.
- Explain procedures for constructing a single-face masonry fireplace.

## Keywords

Air Intake
Ash Pit
Base
Chimney
Chimney Base Flashing
Chimney Cap
Combustion Chamber
Counter-Flashing
Cricket

Draft
Fire Stopping
Firebox
Fireplace Brick
Fireplace Surround
Flue Lining
Hearth
Hearth Base
Inner Hearth

Multi-Face Fireplace
Outer Hearth
Rosin Fireplace
Rumford
Single-Face Fireplace
Smoke Chamber
Smoke Shelf
Throat
Throat Damper

## Chapter Review Questions and Exercises

### SHORT ANSWER

1. What OSHA safety standards are applicable for those working below ground level in potentially entrapping areas between foundation walls and nearby embankments?

    _____
    _____
    _____
    _____

2. Before employers permit their workers to erect, dismantle, or access scaffolds, what are they required to do?

3. When working on scaffolding to build a fireplace chimney, what is closest that one should be to overhead power lines?

4. What can be the hazards of using a rope and pulley attached to scaffolding end frames to elevate masonry materials to workers?

5. What are the potential risks of breathing high levels of masonry dusts such as the dust resulting from power saw operations?

6. As defined by OSHA standards, where is "fall protection" required?

7. What is required for protecting persons from falling to levels 6 feet or more below open areas, such as those existing between floor framing and the hearth base before the concrete hearth support is placed?

8. Why is it important to have a competent person inspect and clean the interior walls of flue linings venting fireplaces, preferably no longer than 12 months between inspections?

9. What are the design requirements for a concrete footing intended to support a masonry fireplace?

CHAPTER 21 *Masonry Fireplaces* 99

10. What is required for a conventional fireplace to bring an outside air supply to the firebox?

11. For firebox openings larger than 6 square feet, how far must the outer hearth extend beyond the sides of the opening? How far must it extend in front of the opening?

12. If a wood mantle projects more than 1½ inches from the face of a fireplace, what is the minimum distance required between the woodwork and fireplace opening, both at the sides and top of the opening?

13. Because the inner hearth form cannot be removed from below once the concrete cures, of what type of material should it be constructed?

14. What type of cement is required for bedding firebrick when constructing a firebox?

15. What should be the maximum width for joints between firebrick?

16. What fireplace dimension determines the design and dimensions for the sidewalls and back wall of a firebox?

17. Measured from the face of the fireplace surround, what should the minimum depth of a fireplace inner hearth be?

18. What should be the minimum clearance between the outside walls of the fireplace and all combustible materials such as wood framing?

19. For the sides of a firebox, what should be the width of the air space separating the backside of firebrick and the brick walls behind them?

20. To permit proper curing of the mortar, how many days does the Brick Industry Association recommend not building a fire in a newly constructed firebox?

21. What is recommended to be placed between a metal throat damper or a steel angle lintel and adjacent masonry to provide for unobstructed thermal expansion of metals?

22. What is the maximum slope from the vertical for the sidewalls and front walls of smoke chambers constructed of corbelled masonry? When using prefabricated smoke chamber linings?

23. Compared to another part of the fireplace, what must the completed smoke chamber be no higher than?

24. In relationship to the height of combustible ceiling joist structural framing near the outside faces of the chimney, where should the first flue lining above the smoke chamber walls begin?

25. What should be used as bedding mortar between each flue liner section?

# Chapter 22 Brick Masonry Arches

## OBJECTIVES

Upon completion of this chapter, you will be able to:
- Identify the types of brick arches.
- Identify and define the parts of an arch.
- Construct a semi-circular brick arch.

## Keywords

Abutments
Bonded Arch
Camber
Circular Arch
Compression
Creepers
Depth
Extrados
Gauged Brick
Gothic Arch

Horseshoe Arch
Intrados
Jack Arch
Keystone
Major Arch
Minor Arch
Multi-Centered Arch
Segmental Arch
Semi-Circular Arch
Skewback

Soffit
Span
Spring Line
Triangular Arch
Tudor Arch
Unbonded Arch
Venetian Arch
Wood Centering

# Chapter Review Questions and Exercises

## SHORT ANSWER

1. What personal protection equipment should one wear while cutting brick shapes for arches?
   _____
   _____
   _____

2. From whom should one receive training before operating power tools?
   _____
   _____
   _____

3. What is the maximum span for a minor arch? What is the load limit a minor arch is permitted to support?

_____

_____

_____

4. Define a "major arch."

_____

_____

_____

5. What is the recommended "depth" per foot of span for both segmental and semi-circular arches?

_____

_____

_____

## COMPLETION

6. The horizontal distance between the supports at each end of the arched opening is known as the _____.

7. The arch _____ is the height of the brickwork forming the arch.

8. Each masonry unit forming an arch ring is known as a _____.

9. A _____ arch consists of two or more voussoirs bonded vertically to form the depth.

10. Tapered joints between voussoirs should be no larger than _____ inch(es) and no smaller than _____ inch(es).

11. Tapered brick, also called _____ brick, permit having uniform width head joints between voussoirs.

12. Weights or forces above arches create what are known as _____ forces.

13. The center voussoir is called the _____.

14. Walls or piers called _____ support an arch.

15. The _____ of an arch is a measure of its height from the imaginary line at which an arch begins to curve and to the center point of the lower edge of the arch ring.

16. Located at the abutments, the _____ are the surfaces on which the ends of an arch rest.

17. Voussoirs are aligned _____ to the arch forms, meaning that the center point of each voussoir touches the curved arch form.

18. A _____ line helps position each voussoir at its proper angle.

19. Placing wooden _____ _____ or _____ above the wood-centering form and between each voussoir prevents mortar at the bottom of the joints, eliminating hardened, untooled mortar at the underside of the arch, which must be cut out once the wood centering is removed.

20. Referred to as _____, the brick adjacent to the arch voussoirs must be cut at an angle.

21. A temporary arch form should remain for a minimum of _____ days.

22. For arches less than _____ feet, flashing can be installed with an end dam at each end in the first bed joint above the keystone.

23. _____ flashing can be installed in the bed joints above longer arches.

24. The radius for a _____ arch is uniform but the arch is less than a semi-circle.

25. Having a rise equal to or greater than its span, the _____ arch comes to a point at its center.

26. The intrados of a _____ arch appears to be horizontal.

27. Arches forming circles are known as _____ arches.

28. A _____ arch is a circular arch greater than the 180° design of a semicircular arch but less than 360° circular arch.

29. Two straight inclined sides form a _____ arch.

30. A _____ arch is a semicircular arch flanked by a narrower horizontal line of brickwork on both sides.

31. Assume that you are constructing anchored brick veneer walls on a home whose window tops have semi-circular and segmental arch shapes. The voussoirs that form the arch rings project well beyond the window frames, causing them to misalign beyond the face of the wall. What can you do to remedy this problem?

_____

_____

_____

# Chapter 23 Cleaning Brick and Concrete Masonry

## OBJECTIVES

Upon completion of this chapter, you will be able to:

- Identify sources for construction dirt and mortar soiling masonry.
- List measures for preventing dirt-stained and mortar-stained masonry.
- Identify recommended cleaning procedures for specific categories of brick.
- Describe methods for cleaning concrete masonry units.

## Keywords

Bleeding  
Efflorescence  
Muriatic Acid  
Propriety Compounds  
Trisodium Phosphate  
White Scum

## Chapter Review Questions and Exercises

### SHORT ANSWER

1. What risks are workers exposed to when scaffold planks next to walls are removed at the end of the work day to prevent masonry debris and splashing rains from staining masonry walls?

2. To whom should the use or disposal of muriatic acid and other brick cleaning agents be entrusted?

3. What health risks are there for those using muriatic acid or other brick cleaning agents without proper personal protection equipment?

4. What personal protection equipment should be worn by those exposed to muriatic acid or other brick cleaning agents and their fumes?

5. What should be posted to keep others away from the work zone where brick cleaning is in progress?

6. To what environments should muriatic acid and other cleaners having potentially harmful fumes be limited?

7. What should be done to all plants and ground cover that may come into contact with brick cleaning agents?

8. What regulations must one comply with for the storage, use, or disposal of chemical cleaning agents?

9. To whom should the use of pressure washers be limited?

10. What personal protection equipment must those using or in close proximity to pressure washers wear?

11. Besides manufacturers' instructions, what agencies and governing regulations must one comply with for the storage, use, and disposal of masonry cleaners such as muriatic acid and propriety compounds?

CHAPTER 23 Cleaning Brick and Concrete Masonry    **111**

12. What government agency is responsible for enforcing safety and health standards such as those intended to protect workers exposed to the hazards of chemical cleaners?

13. Describe clear water rinsing procedures for removing loosened mortar and expended cleaning agents from brick walls to ensure that chemicals and loosened mortar do not stain surfaces below that which is being cleaned.

14. To prevent chemical reactions with acids, what type of scrapers are recommended to remove particles of hardened mortar from the face of brick walls?

15. What type of buckets or pails should be used to hold brick cleaning solutions containing acids?

16. What is considered to be an alternative method for cleaning brick that may react with chemical agents?

17. What hand-held tool is specifically intended for scraping hardened particles of mortar from the faces of CMU walls?

18. List four sources of soiled or mortar-stained brick.

# 112 Workbook to Accompany Residential Construction Academy: Masonry

## COMPLETION

19. Elevating materials onto _____ _____ prevents stains caused by ground soils contaminating stock piles.

20. Soils having higher iron content result in _____-colored stains.

21. Masonry units absorbing salt-laden ground water contribute to _____-colored efflorescence stains.

22. List five precautionary measures taken to minimize the cleaning of masonry units.

    _____
    _____
    _____
    _____
    _____

23. Allowing the mortar to cure for at least _____ days before cleaning is a recommendation of the Brick Industry Association.

24. The _____ and _____ producers should be consulted for recommended cleaning methods, agents, and procedures.

25. List four materials needing protection from cleaning agents.

    _____
    _____
    _____
    _____

26. Three methods for cleaning new brick are:

    _____
    _____
    _____

27. _____ or _____ buckets or pails are recommended to contain cleaning agents because metal buckets corrode in the presence of some cleaning agents and the chemical reactions between cleaning agents and metal buckets may discolor brick.

28. Diluted hydrochloric acid, known as _____ acid, is limited to cleaning only those brick not susceptible to acid staining.

29. _____ compounds are chemicals containing organic and inorganic acids, wetting agents, and inhibitors for brick susceptible to metallic oxidation staining.

30. Solutions of household _____ and water or _____ and water remove mud and soil.

31. When diluting acid in water, it is recommended to pour the _____ into the _____ so that potentially harmful fumes are minimized.

32. Clean water for rinsing should not exceed pressures of _____ psi and should be applied at the rate of _____ to _____ gallons per minute with pressure washers.

33. Cleaning solutions are applied with hand-pumped low-pressure sprayers or powered pressure washers at _____ psi maximum.

34. High-pressure water streams, defined as those greater than _____ psi, may alter a brick's appearance.

35. _____ is a method of cleaning brick masonry using compressed air and one of several types of abrasive materials.

36. Examples of abrasive cleaning materials include _____, _____, and _____.

37. Regardless of the cleaning method or cleaning agent selected, testing the performance and results on a less _____ section should be performed before widespread use.

38. List two damaging effects of using acids to clean concrete masonry units.

39. List two factors contributing to the emergence of specialized cleaning services for both new and older masonry construction.

40. Assume that you are a masonry contractor. You are obligated by contract to clean a home's newly constructed brick walls but the homebuilder is obligated to supply the cleaning agent. Upon reading the printed information provided by the brick manufacturer, you learn that the cleaner provided by the builder is not recommended for cleaning this particular brick. What should you do?

41. Unknown persons have painted graffiti on a brick wall facade. The building's owner has you come to see the wall and to get your opinion on using sand blasting to remove the graffiti. What other options might you recommend before sand blasting the brick wall and why?

42. Assume you are a masonry contractor. You are present when the brick is delivered to the job site. You notice that the deliverer is about to set brick cubes on ground where rainwater is standing. What would you do?

# Chapter 24 Masonry as a Career

**OBJECTIVES**

Upon completion of this chapter, you will be able to:
- Describe the work performed by brick masons.
- Describe the physical qualifications for doing masonry tasks.
- List possible career benefits for brick masons.
- Explain the career pathways for becoming a brick mason.
- Describe the employment outlook and opportunities for brick masons.
- List items for which a masonry contractor must be knowledgeable.
- Describe factors building a favorable reputation for a masonry contractor.

## Keywords

| | | |
|---|---|---|
| Apprentice | Employee | Masonry Contractor |
| Architectural Concrete | Employee Benefits | Masonry Instructor |
| Architectural Drawing | Employer | Material Safety Data Sheets |
| Certification | Estimate | Profit |
| Change Order | Experience Modification Rating | Project Manager |
| Completed Operations Insurance | Estimator | Scale Drawing |
| Construction Contract | Job Foreman | Self-Employed |
| Construction Manual Log | Job Superintendent | Structural Engineer |
| Contract | Journey-Level Worker | Workers' Compensation |
| Contract Bond | Liability Insurance | Working Drawing |
| Contractor | Licensed Contractor | |
| | Markup | |

## Chapter Review Questions and Exercises

### SHORT ANSWER

1. What government agency regulates the standards required for promoting workers' safety and health?
   _____
   _____
   _____

2. What types of information are recorded in a construction manual log?

3. A(n) _____ is one hired to work for another.

4. Besides brick and block, list three other types of masonry materials that brick masons may use to build walls.

5. List three pathways for beginning training to become a brick mason.

6. List four characteristics that one's success as a masonry contractor is likely to depend upon.

7. List three factors influencing the cost for doing work.

8. List three desirable personality characteristics for having a successful business.

## COMPLETION

9. A _____ contractor is one who is granted permission by government authority to operate a contracting business.

10. A _____ _____ guarantees the payment of all debts pertaining to the construction of a project.

11. Government regulations sometimes require _____ before an individual is permitted to operate certain machinery or to use certain chemicals.

12. General _____ insurance provides insurance coverage for personal injury and damage to the property of others.

13. Depending on the number of employees, a contractor may be required to have workers' _____ for insuring employees in the event of accident, injury, or death.

14. The _____ adjusts workers' compensation premiums on the contractor's record of employee injury frequency and cost.

15. Insurance covering damages that either products or services may cause after the work is completed is called _____ _____ insurance.

16. An _____ is a judgment of construction costs.

17. A construction _____ is a legal document stating the terms and conditions for construction of a specific project.

18. List three items included in a contract.
    _____
    _____
    _____

19. Payment for additional work not stated in the contract requires documentation by means of a _____ order.

20. A _____ is the dollar amount over and beyond material and labor costs needed for business operations.

21. A _____ is the amount of money remaining after all expenses have been met.

22. List three ways profit margins can be used in a business.
    _____
    _____
    _____

23. List three items included in a construction daily log.
    _____
    _____
    _____

24. Emergency and first aid procedures as well as storage and disposal information for every product at a job site are included on the _____ _____ _____ _____.

25. What should a contractor do when existing conditions prevent the completion of a job exhibiting acceptable workmanship meeting building code requirements?
    _____
    _____
    _____

26. A professional _____ is licensed to design structures capable of supporting the weights and forces exerted upon them.

27. A masonry contractor should request _____ printed specifications before contracting to build any structural wall.

28. For complying with governing safety and health standards, what personal protection equipment should those employed in the masonry construction industry have available at job sites?
_____
_____
_____

29. Give examples of governing regulations and standards with which masonry contractors must comply.
_____
_____
_____

# Chapter 25  Safety for Masons

## OBJECTIVES

Upon completion of this chapter, you will be able to:
- Demonstrate the proper clothing and personal protection equipment for masons.
- Maintain a safe work area.
- Describe the safe handling and storage of materials.
- Demonstrate the proper use of tools and equipment.
- Report accidents and safety hazards.

## Keywords

ANSI
Competent Person
Coupling Pin
Dense Industrial 65
Double Insulated
Equipment Grounding Conductor
Ground Fault Circuit Interrupters

Guardrails
Laminated Veneer Lumber
Mid-Rails
Mudsill
NIOSH
OSHA
Personal Protection Equipment
Sawn Planking

Scaffold
Scaffold Platform
Supported Scaffolds
Suspension Scaffolds
Toe Boards

## Chapter Review Questions and Exercises

### COMPLETION

1. Trousers must extend from the _____ to the _____.
2. Shirts must have sleeves _____ inches or longer.
3. A _____ sole, _____-high or higher work shoe is required.
4. Safety glasses bearing the marking _____ are required where there is a potential for eye injuries.
5. Prescription eyeglasses must have _____ lenses and _____ shields are required on both sides of the eyewear.
6. To prevent breathing in non-toxic masonry dusts, one should wear a disposable _____ _____ mask.

7. It is recommended having materials no closer to the project under construction than _____ feet or higher than _____ feet.

8. If a loaded wheelbarrow begins to upset and no one is close, then _____
_____.

9. First, cut a brick band at the _____ of a strap of brick.

10. Finish removing the brick band from a strap of brick by cutting on each side about _____ inches from the bottom of the strap and then bending over the ends.

11. Lift heavier objects by stooping and using the _____ muscles instead of bending over and straining the back.

12. Never use a mason's line that is _____, meaning that it is damaged and may break.

## SHORT ANSWER

13. What federal government agency conducts inspections of and issues citations and proposed penalties for employers covered under the Occupational Safety and Health Act of 1970 for alleged violations of applicable safety and health standards?

14. What is the private organization acting as administrator and coordinator for voluntary safety standards?

15. In what areas are employers to ensure that each affected employee uses protective footwear?

16. What standard must eye protection meet when it is required to be worn when machines or operations present potential eye or face injury from such sources as masonry dusts and mortars?

17. What standards govern the exposure to noises of machinery and power equipment?

18. For what conditions do OSHA standards require the wearing of hard hats?

19. What is the intended purpose for wearing a dust mask?

20. For maximum personal protection from electric shocks or fatal electrocution, what must all 120-volt receptacles on construction sites have to protect those exposed to electrical cords or equipment beyond the connection?

21. According to OSHA standards, unless professional medical services are reasonably close in time and distance to a work site, who shall be available at the work site to render first aid?

22. List seven types of information provided by Material Safety Data Sheets.

23. Who are the only workers authorized to erect, dismantle, or access scaffolds?

24. Are cross braces on tubular welded-frame scaffolds to be used as a means of access or egress?

25. What industry standard must sawn-wood scaffold planks meet?

26. What is the maximum for which scaffold planks are permitted to deflect when loaded?

27. Scaffold planking 10 feet or less in length are to extend no fewer than _____ inches or greater than _____ inches beyond the centerline of its support.

28. To meet the intent of OSHA standards, what must be placed below scaffold frame uprights to help distribute weight?

29. At what working height are accessories including guardrail posts, top rails, mid-rails, and toe board clips required for scaffolds?

30. At what heights are 5-foot-wide supported scaffolds to be restrained from tipping?

31. Above what height must workers on scaffolds be protected from falling?

32. Everyone must wear eye protection whenever _____ is working an enclosed masonry training lab.

33. Gloves should be worn when handling _____ materials.

34. Remove all _____ from the body whenever working to prevent injury or the loss of blood circulation if injured.

35. Keep masonry scraps off a masonry training lab floor and put them in a _____.

36. Sprinkle _____ on wet floor areas in a training lab and set a _____ _____ at the spill for others to see.

37. Never leave _____ or _____ in boards, since their sharp, pointed ends can cause injury.

38. Tools must be examined for possible _____ before each use.

39. When using a steel measuring tape, prevent _____ the fingers by keeping them away from the steel tape.

40. Explain what should be done for someone with a possible eye injury as a result of mortar or other masonry debris.

41. What are some of the harmful effects and health risks for someone overexposed to the ultraviolet rays of the sun?

42. What safe distance must one keep from overhead power lines when handling materials or equipment having the potential to cause electric shock or fatal electrocution?

43. What are the health risks for breathing air contaminated with cement dusts, products containing silica, or sand over a period of time? How can these risks be minimized?

___

44. Why must appropriately sized and approved ventilation systems be provided in such places as masonry training labs or interior masonry construction sites?

___

45. Why should one not wear jewelry when working? Give some of the potential hazards one can incur by wearing jewelry such as finger rings, wrist watches, and necklaces.

___

46. What should you do if another worker or even your employer asks you to do something for which you have not been properly trained to do?

___

# Chapter 26: Working Drawings and Specifications

## OBJECTIVES

Upon completion of this chapter, you will be able to:

- Identify the types of lines, symbols, and abbreviations used for drawings and explain where they may be found on a drawing.
- Define and explain the types of working drawings that may be part of a construction document.
- Define, identify, and explain the purposes of elevation drawings, details, and sections.
- Explain the purpose of presentation drawings and how they differ from working drawings.
- Define that part of the construction documents called "specifications" and explain its purposes and contents.
- Become familiar with interpreting working drawings, noting those parts of each applicable to the masonry construction of the project.

## Keywords

| | | |
|---|---|---|
| Bench Mark | Elevations | Project Representative |
| Border Lines | Extension Lines | Scale Drawing |
| Break Lines | Floor Plans | Section |
| Building Materials Symbols | Foundation Plans | Section Lines |
| Centerlines | Framing Plans | Section Reference Line |
| Change Order | General Conditions | Special Conditions |
| Climate Control Symbols | Hidden Lines | Specifications |
| Construction Details | Leader Lines | Supplementary Conditions |
| Construction Documents | Object Lines | Symbols |
| Dimension Lines | Plot Plans | Topographical Features |
| Drawing Identification Symbols | Plumbing Symbols | Topographical Symbols |
| Electrical Symbols | Presentation Drawings | Working Drawings |

## Chapter Review Questions and Exercises

### COMPLETION

1. _____ lines, such as those representing wall lines, are solid lines representing the outline shape of objects.

2. Used only when necessary details are not deleted, solid _____ lines permit creating drawings requiring less space.

3. Solid, bold lines indicating the outer perimeters for any details intended for a drawing are called _____ lines.

4. Points of known elevation used as a reference in determining other elevations are called _____ marks.

5. Three of the six types of building symbols, marks, letters, characters, figures, or combinations of these representing specific objects are _____, _____, and _____.

6. Scales typically used for making construction drawings are the _____-inch-scale and the _____-inch-scale.

7. Scale drawings interpreted by lines, symbols, and abbreviations are called _____ drawings.

8. _____ plans are scale drawings showing the location of the structure in relation to the site and the property corners and lines.

9. Scale drawings showing the foundation walls for the intended structure are called _____ plans.

10. _____ plans are scale drawings showing the overall length and width of the floor framing, including masonry-related details such as fireplaces, hearth extensions, and interior masonry walls.

11. Construction details for brick paved walkways are shown on _____ plans.

12. Two-dimensional graphic scale representations of the front, sides, and back of a structure showing finished grade, exterior wall coverings, doors, windows, and other details are called _____.

13. Similar to photos, _____ drawings illustrate a structure as one is to view it.

14. Some drawings are made at a larger scale than other working drawings to show greater detail. Such drawings are called _____ details.

15. _____ are a written descriptions of the conditions of the contract.

16. The conditions of a contract upon which the owner and the contractor agree are called the _____ documents.

17. A _____ order is a written change to the original contract agreed upon by the owner, architect, engineer, and contractor.

**CHAPTER 26** *Working Drawings and Specifications* **127**

## SHORT ANSWER

18. How can one determine a dimension if it is not given or otherwise not visible on a set of working drawings?

   _____
   _____
   _____
   _____

19. Assume that you are a masonry contractor hired by a homebuilder to construct CMU foundation walls for a home. The homeowner, the person for whom the homebuilder is building the home, asks you to make changes resulting in wall designs other than those specified in the plans. What should you do?

   _____
   _____
   _____
   _____